Electrical Installation Designs

Also of interest

*A Handbook on the 16th Edition
of the IEE Regulations
for Electrical Installations*
ECA of Scotland, ECA & NICEIC

A Practical Guide to the Wiring Regulations
Geoffrey Stokes

Electrical Safety and the Law
A Guide to Compliance
Second Edition
Ken Oldham Smith

Electrical Installation Calculations
B.D. Jenkins

Electrical Distribution in Buildings
C. Dennis Poole
Second Editions revised by
Trevor E. Marks

Electrical Installation Designs

Bill Atkinson

OXFORD

BLACKWELL SCIENTIFIC PUBLICATIONS

LONDON EDINBURGH BOSTON

MELBOURNE PARIS BERLIN VIENNA

© 1994
Blackwell Scientific Publications
Editorial Offices:
Osney Mead, Oxford OX2 0EL
25 John Street, London WC1N 2BL
23 Ainslie Place, Edinburgh EH3 6AJ
238 Main Street, Cambridge,
  Massachusetts 02142, USA
54 University Street, Carlton,
  Victoria 3053, Australia

Other Editorial Offices:
Librairie Arnette SA
1, rue de Lille
75007 Paris
France

Blackwell Wissenschafts-Verlag GmbH
Düsseldorfer Str. 38
D-10707 Berlin
Germany

Blackwell MZV
Feldgasse 13
A-1238 Wien
Austria

First published 1994

Set by DP Photosetting, Aylesbury, Bucks
Printed and bound in Great Britain by
Hartnolls Ltd, Bodmin, Cornwall

DISTRIBUTORS

Marston Book Services Ltd
PO Box 87
Oxford OX2 0DT
(*Orders*: Tel: 0865 791155
         Fax: 0865 791927
         Telex: 837515)

USA
Blackwell Scientific Publications, Inc.
238 Main Street
Cambridge, MA 02142
(*Orders*: Tel: 800 759-6102
                 617 876 7000)

Canada
Oxford University Press
70 Wynford Drive
Don Mills
Ontario M3C 1J9
(*Orders*: Tel: (416) 441-2941)

Australia
Blackwell Scientific Publications Pty Ltd
54 University Street
Carlton, Victoria 3053
(*Orders*: Tel: 03 347-5552)

British Library
Cataloguing in Publication Data
A Catalogue record for this book is available from
the British Library

ISBN 0–632–03703–2

Library of Congress
Cataloging in Publication Data
Atkinson, Bill.
    Electrical installation designs/Bill Atkinson.
        p.   cm.
    Includes index.
    ISBN 0-632-03703-2
    1. Electric wiring—Interior.   2. Electric
    engineering.
    I. Title.
    TK3271.A8   1994
    621.319′24—dc20   93-27585
                                         CIP

Contents

Preface

There are many books on electrical installation practice where the focus is on calculations and regulations. *Electrical Installation Designs* has been written from a different viewpoint. Typical projects are examined to produce designs that will fit current standards.

Most electrical contractors have an understanding of requirements related to their own regular everyday activities. Work is carried out using rule-of-thumb methods. Repetitive designs are used. Many installers claim that they are not designers and show concern that they are now required to certify the adequacy of an installation design.

In practice problems only arise when an unusual project is undertaken or there is a change in regulations.

There is no harm in using a standardized design, rather in the way that an experienced cook uses a published recipe for a cake. *Electrical Installation Designs* is a book of recipes. The installer may select a design that corresponds as near as possible to the contract in hand and take up such technical and regulatory advice as is required. This will reduce lengthy calculations and detailed study of the 74 chapters in the 16th edition of the Wiring Regulations.

Most basic electrical installations may be completed by a competent person with appropriate guidance to avoid serious problems and hazards.

Project chapters illustrate methods that could be used for particular types of installation ranging from a house to an industrial workshop. The ideas are by no means exclusive. Alternative solutions are always possible. In many instances detailed calculations and different circuitry will be more profitable. By their very nature, simplified examples of fictional projects can only produce generalized results.

It is known that many contractors appreciate the monthly down-to-earth interpretations of the Wiring Regulations given by Bill Atkinson in the *Electrical Contractor* journal. This approach is taken up in the special topic chapters in this book. Earthing, switching and overcurrent protection rules are explained.

Electrical installation students and non-electrical associates in the construction industry will appreciate the user-friendly approach. Nevertheless, this is not a do-it-yourself book for the untrained person. Warnings are given where more specialized study is necessary. For example, readers are advised not to embark on flameproof installations without further training. Apart from moral implications and contractual risks, statutory requirements are such that incompetent work may carry criminal penalties.

Although the emphasis is on tried and tested methods, some new techniques are introduced. Most significant is the option for tree circuitry as an alternative to outdated ring mains. This is the first book to give designers the opportunity to compare the advantages of the tree system for both domestic and commercial installations. In recent years consumer requirements have changed. It is essential that the industry keeps an open mind on changes in traditional wiring practice.

A.G. Smith

Chapter 1

Introduction

This book contains designs for electrical installations which have been prepared with reference to Wiring Regulations and there are interpretations of particular technicalities.

This is not a do-it-yourself book for the amateur or untrained person. It is a guidance manual for competent electrical designers and students of installation practice.

As far as possible all information accords with the requirements of the 16th edition of the IEE Wiring Regulations under its new designation as BS 7671. Relevant Regulation numbers and other references are shown in the margin. (Because of the space allowance the following abbreviations have been used: Ch. – Chapter; Sec. – Section; Defs – Definitions; App. – Appendix.) Reference is also made to various other British Standards and related Health and Safety documentation.

Layout of chapters

Interspersed throughout the book are two types of chapter giving information in different formats.

- *Project chapters*. These may be compared with a selection of recipes for an experienced chef. The recipes give ideas for the design of typical electrical installations. Each project is dealt with on a stand-alone basis. Cross reference between these chapters is avoided and similar information may be found for more than one scheme.
- *Topic chapters*. These supplement the project chapters with in-depth discussion of generalized technicalities. They also provide study information on regulatory subjects. It may be necessary to refer to these details to finalise a design with particular problems.

Wiring Regulations

Throughout this book the terms Wiring Regulations (or Regulations) refer to the 16th edition of the Wiring Regulations issued by the Institution of

Electrical Engineers (now BS 7671). The Standard therefore represents a code of acceptable safety for electrical installations to protect:

    ❑ Persons,
120–01 ❑ Property, and
    ❑ Livestock,

against electrical hazards which are described as:

    ❑ Electric shock,
    ❑ Fire,
    ❑ Burns, and
    ❑ Injury from mechanical movement of electrically actuated machinery.

The Regulations are not a statutory document but are quoted as a means of
120–02 compliance with certain statutory instruments. It would appear that a criminal charge could not be brought for failure to comply with the Wiring Regulations, but such failure could be used in evidence on a charge for breach of the Electricity Supply Regulations or the Electricity at Work Regulations.

It would be most unwise to ignore any of the requirements of the Regulations. They must be considered in their entirety and are a *pass* or *fail* test. An installation cannot partially comply.

Scotland

Different considerations apply in Scotland where the Wiring Regulations are quoted as a means of compliance with Building Regulations (Scotland).

Terminology

In order to understand technicalities, the importance of correct terminology is stressed throughout the book. In general, however, the use of over-complicated expressions and trade jargon has been avoided.

The Wiring Regulations carry a list of definitions for words and expressions which may not accord with standard dictionary definitions. Wherever there is any doubt, the Wiring Regulations definition should be applied.

Competence and responsibility

Any person involved with the installation of wiring in buildings takes on both legal and moral responsibilities for safety. A high level of technical and practical competence is essential. This can only be achieved with the appropriate study.

There are always three components to an electrical installation project:

❏ Design,
❏ Installation,
❏ Inspection and test.

Often one person or company takes on all three responsibilities, especially for simple repetitive jobs such as house wiring. On larger schemes, specialist companies may be contractually involved for each aspect and in turn use a team of operators. As the work progresses from planning to completion there must always be one or more supervising individuals who will eventually certify that the three aspects of the contract have been carried out in accordance with the Wiring Regulations and any other statutory or specification requirements.

Procedures

Design

It is sometimes thought that the use of tried and tested methods removes the design aspect from a scheme. This is not the case. Every project involves electrotechnical design decisions which are not to be confused with architectural or customer instructions for the physical location of electrical equipment. Thus, a self-employed electrical contractor who makes a decision on the selection and connection of an electrical accessory is a designer. The same applies to an electrician who makes a similar decision on behalf of an employer.

514-09 All technical design information must be recorded. This is a Wiring Regulations requirement. IEE guidance is that it is essential to prepare a full specification prior to commencement or alteration of an electrical installation. The size and content of the specification will correspond with the complexity of the work. For simple jobs a few lines may suffice.

The designs shown in the following chapters are for guidance and each one includes a suggestion for a suitable design specification. A person selecting this guidance makes a design decision and therefore becomes the responsible person.

742-02 Upon completion of the contract the designer certifies that the design work is in accordance with the Wiring Regulations.

Installation

Where a technical design is drawn up by an electrical engineer or other competent person, it should not be the installer's job to check design details, unless this is one of the contract requirements. The installer is always under an obligation to point out to the designer any obvious conflict with regulations or standards and an installer should always refuse to carry out substandard work.

There would be no defence in law against creating an unsafe installation on the basis of inherently bad instructions.

The installer will use the designer's specification document as required by the Wiring Regulations. This may only cover performance requirements or may give full technical details for the selection and erection of equipment. Once again it must be emphasized that a non-technical instruction to take an electrical supply to a particular appliance or location does not constitute design information.

The installer has the responsibility to ensure that equipment is installed correctly and in accordance with the specification, supplemented by manufacturers' information. The installer is often delegated other tasks such as that of negotiating with the electricity supply company and verifying local licensing requirements.

Upon completion of the project, the installer certifies that the installation work has been carried out in accordance with the Wiring Regulations.

Inspection and test

711–01

No matter how simple or straightforward the job, test procedures must be carried out both during the course of the work and upon completion. This applies equally to work carried out by a single self-employed operator. Self-certification is normally acceptable provided that the contractor has the competence and equipment to test correctly. The customer or an insurer may require specialist certification. This applies more particularly in the case of safety alarm systems or work in hazardous areas.

Whether an in-house or independent specialist, the inspector must be given the full design documentation with amendments showing any relevant on-site modifications. On larger projects this will include 'as fitted' drawings.

Certain parts of the installation may be hidden from view upon completion. In such cases the inspector must arrange for inspection during the course of erection or receive certified confirmation that the work is satisfactory.

741–01

Upon completion of the project, the inspector certifies that the inspection and test has been carried out in accordance with the Wiring Regulations.

Completion

742–02

The signatures of the designer, installer and inspector are required for the Completion and Inspection Certificate. This cannot be issued until the work has been completed in accordance with the Wiring Regulations. Where there are acceptable departures from the Regulations, these must be shown on the certificate.

See Chapter 15 for inspection and test procedures.

Working methods and materials

130–01 The Regulations require that good workmanship and proper materials shall be used.

Operatives

Any person carrying out electrical work must be competent, trained and skilled in the type of installation work being carried out. Where trainees or unskilled operatives are employed for electrical work they must be appropriately supervised.

Workmanship must be of a quality appropriate to the location. A working knowledge of the building structure is necessary where holes and fixings are made to carry cables. Decor should be disturbed as little as possible with prearranged responsibility for making good.

Materials

The Regulations require that every item of equipment shall comply with a
511–01 British Standard or harmonized European Standard. Alternatively, equipment complying with a relevant foreign standard may be used provided that the designer confirms that the equipment provides a degree of safety acceptable to the Regulations. This may mean product certification by an approvals organization.

Chapter 2

Three Bedroom House

At one time, domestic electrical installations were simple and only basic design planning was necessary. A good electrician could be sent on site with a van load of wiring materials with no written instructions or drawings. The installation arrangements were rule-of-thumb and the quality of the job depended on the craftsmanship of the operative.

Any special requirements or missing information could be negotiated on site. Costing was repetitive and easy. The contract price was a simple multiple of the number of lights and sockets.

Times have changed. There is probably no such thing as an average householder. Most occupiers have specialist requirements based upon the choice of room utilization, decor, hobbies and the activities of the various residents.

It is not easy for an architect to forecast the furniture layout in a room. A modern speculative electrical installation cannot make universal provision for every conceivable arrangement. The IEE Guidance Notes suggest that a project should be discussed with the client. This is essential for a custom-built house. As an alternative the installation could incorporate some design flexibility so that the new family is not restricted to bed positions or where kitchen equipment may be plugged in.

Standards for the house industry are determined by the National House Building Council (NHBC). Most building societies and other mortgage lenders require compliance with NHBC requirements.

This chapter starts by illustrating a basic, cost-conscious electrical installation. A scheme may be lifted straight from the pages for such a contract. For more advanced schemes it is hoped that developers will be enticed into better electrical facilities with a 'modern living' theme. A good quotation will include optional extras for improved lighting and socket-outlet facilities. Not all house purchasers want the cheapest possible electrical installation.

The bare minimum

The following is an outline of basic requirements for a three-bedroom house

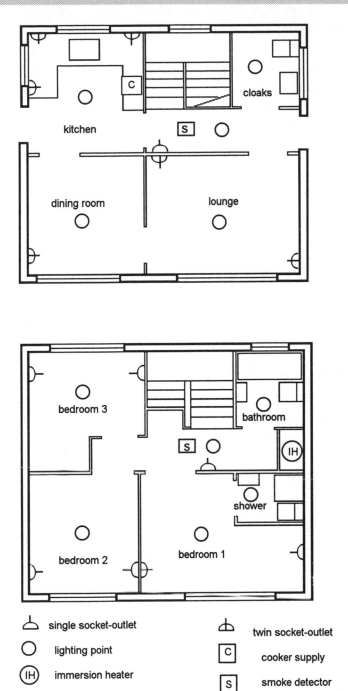

Figure 2.1 Typical three bedroom house.

with 120 m² floor area. This would be a typical speculative estate development (see Fig. 2.1):

❏ *Rooms*
Small kitchen
Dining room
Lounge
Downstairs cloakroom
Main bedroom with en-suite bathroom
Second double bedroom
Small single bedroom
Landing bathroom or shower room
❏ *Heating*
Central heating by gas or oil
❏ *Garden*
There is a small garden at both front and rear

Standards

National House Building Council (NHBC)

The NHBC gives minimum standards for living accommodation and services. The electrical requirements are shown in Table 2.1.

Relevant IEE Wiring Regulations

13 A socket outlets

471–16   ❏ Any socket outlet which may reasonably be expected to supply portable equipment for use outdoors must be provided with 30 mA rcd supplementary protection.

❏ No 240 V sockets are permissible in bathrooms or shower rooms.

601–10   ❏ Where a shower cubicle is located in a room other than a bathroom, e.g. a bedroom, any 240 V socket must be at least 2.5 m from the shower cubicle.

Lighting

314–01   ❏ To avoid danger and inconvenience there should be more than one lighting circuit.

601–08   ❏ In a bathroom or shower room the light must be operated by a pull cord switch. (Alternatively the light switch may be outside the bathroom door.)

601-11 ❐ In a bathroom or shower room a lampholder within 2.5 m of the bath or shower cubicle must be shrouded in insulating material.

❐ A batten holder must be of the type with a protective skirt.

Table 2.1 **NHBC electrical requirements for a house.**		
13 A socket outlets (twin sockets count as two outlets)		
Room	*Outlets*	*Notes*
Kitchen/Utility	6	Where homes have separate areas, the kitchen should have a minimum of four outlets and the utility room two. Where appliances are provided, at least three outlets should be for general use.
Dining room	2	
Living room	4	At least one double outlet family room should be near the TV aerial outlet.
Bedroom	3(2)	Three for main bedroom. Two for other bedrooms.
Landing	1	
Hall	1	
Combined rooms should have sockets equal to the sum of the number for individual rooms with a minimum of seven in the case of kitchen/utility and another room.		
Lighting		
Every room should have at least one lighting point. Two-way switching should be provided to staircases.		
Smoke detectors		
For this two storey house two mains operated, interconnected alarms are required.		

Building Regulations

Smoke detectors

All domestic dwellings, including conversions, must either have:

❑   A complete British Standard fire alarm system, or

❑   Mains operated smoke alarms, one on each floor and interconnected.

Load assessment

311–01   The Regulations require that the characteristics of the supply, including an assessment of the maximum demand, should be determined by calculation, measurement, enquiry or inspection.

It is assumed that this house will be connected to an underground PME 240 V a.c. supply and that the maximum demand load will be less than 100 A. The other characteristics may be determined by enquiry to the electricity company.

A typical domestic supply

Except in unusual circumstances supply characteristics are:

313–01   ❑   *Prospective short circuit current at the origin*
          Never more than 16 kA and most likely less than 2 kA.

❑   *External earth fault loop impedance Z_e*
    Not exceeding 0.35 ohms and most likely less than 0.2 ohms.

❑   *Main fuse*
    This will be to the usual electricity company standard,
    BS 1361 Part 2 or BS 88 Part 2 or Part 6. 100 A.

Provided that these figures apply, there are no problems in applying a standardized electrical design.

Project specification

It is necessary to produce a Project Specification as in Fig. 2.2. This will be used initially for pricing purposes. It will eventually be updated to form the basis of a user manual.

Wiring systems and cable sizes

Circuit design is discussed in detail in Chapter 9. Traditionally domestic installations have been wired using the 'three plate rose' lighting system and ring circuits for power socket outlets.

The specification illustrated for this three bedroom house uses *conventional circuits* as described in the *IEE On-Site Guide*. The Guide explains

Project Specification
BS 7671

Name..................... Location.................

Reference................ Date.....................

240 V 50 Hz. TN-C-S. Supply fuse 100 A BS 1361 or BS 88

PFC less than 16 kA. Earth loop impedance less than 0.35

Consumer unit   BS 5486. Split bus-bar

100 A main switch   63 A 30 mA rcd
4 + 4 way   M6 type 2 mcbs

Circuits	Rating (A)	Cable size (mm^2)	Max length (m)	Lights/points g = gang
1. Cooker	30	6.0	43	1
2. Ring 1 upstairs	30	2.5	71	...1g ...2g
3. Ring 2 downstairs	30	2.5	71	...1g ...2g
4. Ring kitchen	30	2.5	71	...1g ...2g
5. Immersion Heater	15	2.5	35	1
6. Lights upstairs) Bathroom fans)	5	1.0	43	
7. Lights downstairs	5	1.0	43	
8. Boiler	5	1.0	43	
9. Smoke detectors	5	1.0	43	

Figure 2.2 **Project specification for standard three bedroom house.**

that this system will comply with the Regulations and, provided that the circuit cable lengths are not exceeded, no calculations are necessary.

Lighting

Various arrangements for lighting circuits have become standardized in different localities. Two methods are shown in Fig. 2.3 and these utilize twin and earth cables for all runs except two-way switch linkages. It is important to 512–05 run conductors in pairs now that a European Directive requires the reduction of electromagnetic interference which may be a particular nuisance with fluorescent lighting or where dimmer-switches are used.

Wiring should be arranged with phase/neutral or feed/return cables twinned to minimize interference. Separate single-core cable runs should be avoided. For the same reason the most suitable two-way switching arrangement is as shown in Fig. 2.4.

It is a matter of choice whether the connection on to the next lighting point is made at the wall switch or ceiling rose. As far as possible the systems should

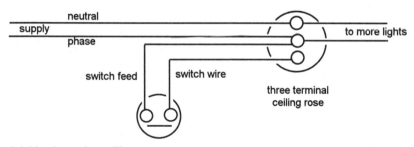

(a) Live looped at ceiling rose.

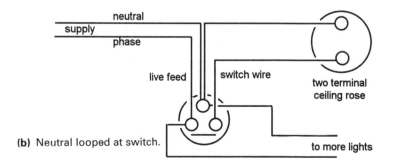

(b) Neutral looped at switch.

Figure 2.3 **Alternative lighting circuitry using twin cable.**

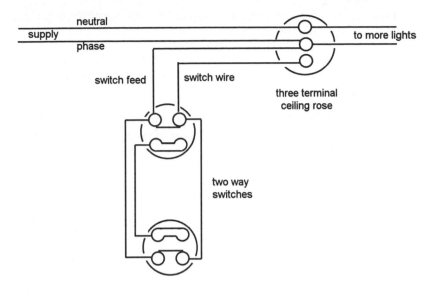

Figure 2.4   **Two way switching avoiding inductive problems.**

not be mixed. This will only cause confusion when alterations or periodic testing are undertaken.

13 A socket outlets

There are several reasons (see Chapter 9) why a ring circuit is not always the best way to service sockets. This design shows traditional arrangements with three ring circuits but alternative 'tree' circuits may be considered to be more appropriate and save wiring.

Cable sizes

Conventions on sizing have developed in the industry. These are the sizes shown on the Project Specification and are adequate for lengths of run as indicated. For very large houses it is generally more economic to run one or more sub-mains to remote areas, rather than increasing final circuit cable sizes.

Circuit protection

Rewirable fuses

533–01   Most people in the electrical industry would like to see the demise of rewirable fuses. The Regulations express a preference for other devices but they are still permissible. Over the years they have proved to be very reliable and their use is likely to continue indefinitely.

◻   *The biggest asset* of a rewirable fuse is its fail-safe characteristic.
◻   *Disadvantages* are inconvenience and abuse by incompetent house-holders.

Despite the above-mentioned drawbacks, it has to be remembered that fuses rarely blow and more sophisticated protection methods are comparitively expensive and have other disadvantages.

Cartridge fuses

341–01 ◻   Nobody ever has a spare cartridge fuse and makeshift repairs can be more hazardous than that on a rewirable fuse.

Miniature circuit breakers (mcb)

◻   An mcb is very expensive for such infrequent use but it is the most user-friendly device when there is a fault.

Earth leakage protection

471–16 To a significant extent the overcurrent protective device debate has been solved by a new requirement for shock protection in the 16th edition of the Wiring Regulations. This is that any socket outlet rated at 32 A or less which may reasonably be expected to supply portable equipment for use outdoors shall be provided with supplementary protection, i.e. a 30 mA rcd.

For this particular design project it would be difficult to assess which of the domestic sockets come into this outdoor-use category. Certainly all the downstairs sockets are potential lawnmower connections. Therefore, for peace of mind and in the interests of standardization a decision has to be taken. It is recommended that the norm for all domestic installations should be for each socket outlet to be rcd protected. Various methods are considered in Chapter 7, but the result will inevitably be associated with an mcb consumer unit.

Arrangement of circuits

Residual current protection

It has been shown above that all socket outlets which could be used to supply portable equipment outdoors must have 30 mA protection. All sockets on this

particular design will be protected by an rcd. Nevertheless, it would be
314–01 inconvenient and cause nuisance tripping if the whole house was supplied
through one very sensitive 30 mA rcd.

There is no requirement for rcd protection for lighting circuits or those
supplying fixed heaters such as the cooker or immersion heater. In the latter
situations shock-protection is achieved by good earthing and low earth loop
413–02 impedance. The rcd protection may be confined to socket outlet circuits.

Miniature circuit breakers

Although rewirable fuses are acceptable, the involvement of fuseways with
and without rcd protection would require an untidy arrangement of enclosures
and cable tails. For this reason, and no other, this standard specification will
be restricted to miniature circuit-breakers.

On a lighting circuit, a type 1 mcb may sometimes cause minor irritation by
tripping when a lamp fails. The selection of mcb types is covered in Chapter
4. For this domestic installation types 2 or B would be appropriate, with
virtually any M rating in a 16 kA conditionally rated enclosure.

Split load consumer unit

There are many ways of making the separation:

❐   *Circuits with rcd protection*
    All 13 A socket circuits
❐   *Circuits without rcd protection*
    Lighting
    Immersion heater
    Cooker
    Boiler
    Smoke detectors

In this case, probably the most cost-effective consumer unit arrangement is to
install a board with provision for ten outgoing circuits plus room for an rcd.
The number of outgoing ways on the specification will depend upon the size
of the rcd module (see Fig. 2.5).

Main switch

The purchaser of a single consumer unit has little choice in the selection of the
main switch. This will be specified by the manufacturer. Most units will
accommodate a 100 A double pole isolator. A switch with a lower current
rating is inappropriate for a whole-house load.

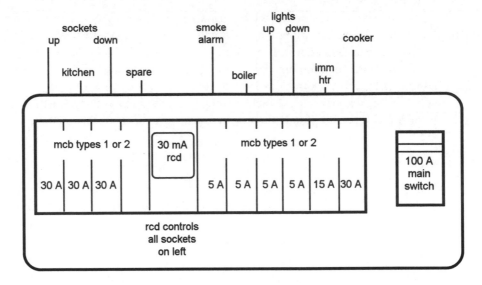

Figure 2.5   **Main switch controls all circuits.**

Earthing and bonding

542–04   A main earthing terminal (MET) must be established. This connects:

❑   The 16 mm^2 supply earthing conductor.
❑   10 mm^2 main bonds to water, gas or oil services.
❑   10 mm^2 circuit protective conductor to the distribution board.

547–02   All cable sizes are the minimum required for compliance with PME Regulations. It should be possible to remove any of the conductors for testing without disturbing others.

   The MET should preferably be external to the consumer unit but in any case it must be easily accessible for disconnecting the supply earth for testing purposes.

   If there is more than one consumer unit or distribution board the MET must carry all bonding conductors and should be externally accessible. Separate circuit protective conductors should then be taken to the earthing bar within each board (see Fig. 2.6).

Gas bonding and external meters

The Wiring Regulations require that bonding to the gas supply should, wherever possible, be within 600 mm of the point of entry into the house and before any branch pipework. This has often proved to be difficult where the

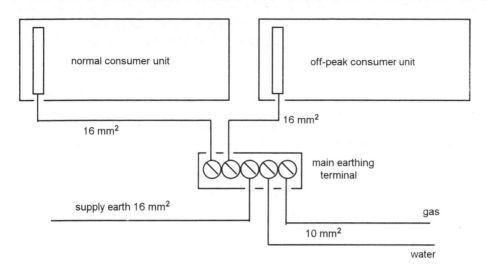

Figure 2.6 **Earthing connections to two services.**

gas pipe is buried in the floor. The only solution is to make the bond at the earliest point where the gas pipe surfaces.

Meters for new gas supplies are now being located in what are known as semi-concealed meter boxes. It is advisable to make the gas bond within this box. British Gas is making provision for the connection. There is a cable hole in the side of the meter box, close to the gas-pipe entry. Within the box, an earth-tag washer is provided at the outlet adaptor.

The 10 mm^2 bonding conductor should be taken through this hole and connected to the washer with a crimped lug and the appropriate label fitted. The cable should be as short as practicable with no spare length curled up. It must not be taken through the gas-pipe hole.

Supplementary bonding

601–04 The only requirement for supplementary bonding is within the bathroom or shower room. For convenience consider a bathroom 'zone'. This may include the airing cupboard if it is within the immediate vicinity. Often two bathrooms back on to each other or the airing cupboard.

543–03 The Wiring Regulations require bonding connections to be accessible for inspection and testing. They are not required to be permanently visible or obtrusive. Connections within the airing cupboard are preferable, or under the bath, or within a pipe duct. In such locations arrangements for access must be provided even if this requires screws to be taken out and bath panels removed.

The purpose of supplementary bonding is to ensure an equipotential situation within this zone. It is an internal cross-bonding arrangement. There is no requirement for a cable to be taken out of the zone, for example back to the mains position.

A convenient reference point for all bathroom/shower room supplementary bonding is the cold water riser in the airing cupboard. If this is copper, it will be in direct contact with the main stop-cock bond.

601–04   Supplementary bonding connections can be taken from this reference pipe to central heating pipes within the airing cupboard. This cross-bonding should be adequate for the bathroom zone (see Fig. 2.7).

There is no requirement to bond bath or shower grab-handles, rails, windows or any other isolated metalwork. It could even be said that any such bonding introduces a potential hazard. However, it may be a wise precaution to test these isolated rails to ensure that fixing screws have not punctured cables within the wall.

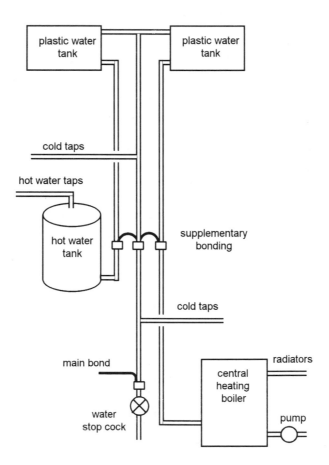

Figure 2.7   **Supplementary bonding in airing cupboard.**

Chapter 3
A Block of Retirement Flatlets

Flats and maisonettes are built in large and small complexes with a range of floor levels at both high and low rise. The pattern of electrical distribution varies in accordance with the developer's arrangements for metering and the electricity company's distribution system.

An example is given for a block of 11 retirement flatlets with some common facilities provided by the landlord. This may be local authority rented accommodation or a private scheme for sheltered housing.

The basic electrical installation specification could be adapted for student accommodation or self-catering holiday flats.

Two schemes

There are two components to the electrical design which will be treated separately:

☐ *Part 1*. Tenants' installations and wiring within flats,
☐ *Part 2*. Landlord's installation and services in common areas.

Early considerations

Metering and distribution

This subject requires early discussion with the electricity company and the client. The whole electrical distribution system depends upon who pays the electricity bills and where meters are to be sited.

Metering alternatives include:

☐ One landlord's metered supply with either unmetered services to tenants or landlord's metering within flatlets,
☐ Individual electricity company services to tenants with meters in flats,
☐ Individual electricity company services to tenants with central metering.

The last arrangement is usually preferred by the electricity company.

There are advantages with centralized meter reading and having the landlord take responsibility for electrical distribution in the building. A disadvantage may be the possibility of vandalism at the central meter room.

This exercise uses the above central metering system. The plan provides the opportunity to design distribution mains to flats and to answer common problems with earthing arrangements for multiple dwellings.

Other interested parties

- ❏ *Fire authority*
     Fire alarms and emergency lighting
- ❏ *Environmental health authority*
     Landlord's kitchen and common rooms
- ❏ *Lift installer*
     Special requirements
- ❏ *Aerial specialist*
     Amplifier for TV distribution system
- ❏ British Telecom

Building details

Construction

This is a new development, but the electrical scheme could be adapted for a refurbishment or conversion contract.

- ❏ *Design* (see Fig. 3.1)
     Three storey, four flats per floor on the upper two levels, three flats plus common rooms on the ground floor.
- ❏ *Walls*
     Externally and between flats, brick or other masonry.
     Partition walls within flats, plasterboard on timber studding.
     Plastered internally.
- ❏ *Floors*
     Ground floor, heavy concrete base, upper floors concrete beams all with 50 mm levelling screed.
- ❏ *Ceilings*
     Ground and first floor, plasterboard on battens.
     Second floor, plasterboard on timber beams under a pitched roof.
- ❏ *Lift*
     Hydraulic mechanism.
- ❏ *Heating*
     Gas or oil central boiler in each flat and one in the landlord's common room area.

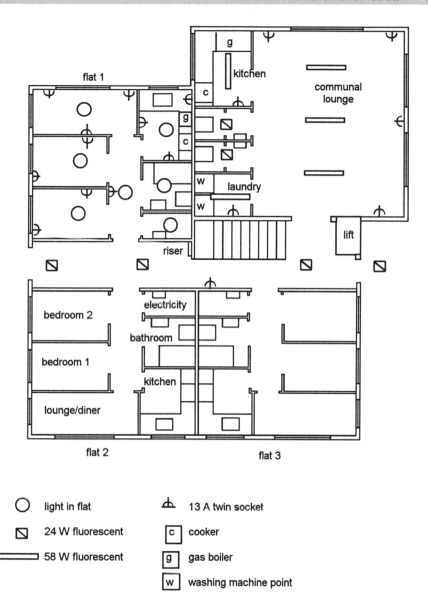

flat 1

kitchen

communal
lounge

c

g

c

w

w laundry

lift

riser

bedroom 2

electricity

bathroom

bedroom 1

kitchen

lounge/diner

flat 2

flat 3

○	light in flat	⟂	13 A twin socket
◪	24 W fluorescent	c	cooker
▭	58 W fluorescent	g	gas boiler
		w	washing machine point

Figure 3.1   **Layout of ground floor flats.**

PART 1 – FLATS

Mains distribution

Main switch fuse

For convenience the mains distribution system and wiring is dealt with in Part 2 of this chapter, which covers the landlord's areas.

The electricity company's supply tails will connect from the meter into a 63 A switch fuse at the central metering position (see Fig. 3.2). Each consumer's supply will be taken through the building to a consumer unit within each flat.

Electrical requirements in flats

All flats are similar, with a conventional installation. The number of outlets is typical and broadly based upon NHBC standards. The fact that we are considering retirement dwellings may encourage the developer to be more generous with outlets than Table 3.1.

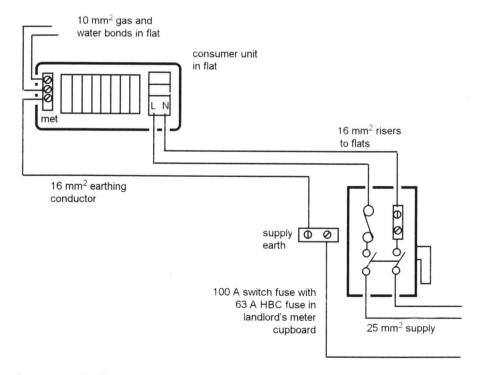

Figure 3.2   **Distribution to flats.**

Table 3.1 **Basic electrical requirements for flats.**

Rooms	Lights	13 A sockets (g = gang)		Other
		1 g	2 g	
Kitchen	1		3	Cooker
Lounge/diner	1	—	3	
Bedroom 1	1	—	2	
Bedroom 2	1	—	2	
Bathroom	1	—	—	
Hallway	1		1	Doorbell Smoke detector

The format for a Project Specification for a development of this type is shown in Fig. 3.3. This layout could be used for various schemes with full details completed to suit the work in hand.

Load assessment and maximum demand

311–01 There is no need to make any detailed assessment. The electricity company will give a standard 100 A domestic service.

Wiring system

Cables will be contained within the concrete floor and lightweight walls. A
522–06 rigid plastic conduit system would be preferable, using unsheathed single-core cables, but difficulty will be encountered in making joints between floor and wall conduits.

There is no Wiring Regulations requirement for rewirability but if this feature is specified, a proprietary flexible conduit system could be used. A typical system can be buried in concrete screeds and utilises single core cable with special outlet boxes and jointing arrangements.

Probably the best 'off the shelf' solution, without rewirability, is to use standard twin-and-earth cable. Theoretically, twin-and-earth cable may be buried direct into concrete but for wiring convenience, plastic conduit will be cast into the floor screed.

Wiring hints

❑ Light gauge, rigid PVC conduit is suitable for casting into concrete screeds but care must be taken to prevent damage to the conduit before

Project Specification BS 7671				
Name....................... .Flat No. .... Location....................... Reference................ Date.....................				
240 V 50 Hz. TN-C-S. Supply fuse 100 A BS 1361 or BS 88 PFC less than 16 kA. Earth loop impedance less than 0.35				
Main switch-fuse in meter cupboard 63 A BS 88 fuse Sub main to flat 16 mm^2 swa 3-core				
Consumer unit in flat BS 5486. 100 A main switch 6 way M6 Type 2 mcbs				
Circuits	Rating (A)	Cable size (mm^2)	Max length (m)	Lights/points g = gang
1. Cooker	32	6.0	43	1
2. Ring 1 kitchen) Boiler)	32	2.5	71	3 x 2g
3. Ring 2	32	2.5	71	7 x 2g
4. Lights 1) Bathroom fan)	6	1.0	43	3
5. Lights 2	6	1.0	43	4
6. Smoke detector	6	1.0	43	

Figure 3.3 **Project specification for a typical flat.**

the concrete is poured. Heavy gauge conduit is more robust but has slightly less cable space.

❏ It is essential to have at least 16 mm of concrete cover over plastic conduit. This may be difficult if there are crossovers in the conduit runs.

❏ Ensure that there are draw wires or strings left in all unwired buried conduits.

Wiring in false ceilings

On the ground and first floors the ceilings are of plasterboard fixed to timber battens on the soffit. The use of sheathed cable is acceptable in this space but subject to the same protective requirements as those for cables under floor-boards. The Wiring Regulations require that unprotected cables which do not incorporate a metallic sheath, when installed above a plasterboard false ceiling, must be at least 50 mm from the underside of the batten.

522–06

A convenient arrangement to comply with this requirement is to cross-batten the ceiling to give a 75 mm space above the plasterboard. This has the added advantage that cables may be easily routed in any direction with minimal drilling or notching.

Cross-battening arrangements should be negotiated early in the contract. Where there are on-site objections from the builder, cables must either be mics; or PVC installed in earthed conduit; or given equivalent physical protection against penetration by nails. This is obviously an expensive alternative to the cross-battening or a notched 75 mm batten.

The use of thin metallic or plastic cable capping as physical protection is not adequate in these locations.

Wiring in roof space

This may be conventional twin-and earth cable fixed to the timber joists. The roof space will get hot in summer months and it is advisable to keep cables clear of thermal insulation. The lighting loads for flats is minimal and no cable derating is necessary.

523–04

Cable sizes

All cable loads are relatively low and distances are short. No special factors apply (see Table 3.2).

App. 4

Table 3.2 Cable sizes and circuitry for flats.		
Circuits	*mcb rating* (Type 1 or 2)	*cable size* (mm²)
Lighting 1	6 A	1.0
Lighting 2	6 A	1.0
Smoke detector	6 A	1.0
Sockets 1	20 A tree	2.5
	or	
	32 A ring	2.5
Sockets 2	32 A tree	4.0
	or	
	32 A ring	2.5
Cooker	32 A	6.0

Arrangement of circuits

Lighting

314–01   Even though these are small flats, there must be more than one lighting circuit per dwelling to ensure that the operation of a single device does not plunge the flat into darkness. This applies to all types of dwelling with more than about four lights and is especially important for sheltered accommodation.

Smoke detector

A separate 6 A circuit is required to comply with the Building Regulations.

Socket outlets

The kitchen should be considered separately. Either a 30 A ring or 30 A tree system is appropriate for heavy loading. For the rest of the flat a 20 A tree system would be acceptable or a 30 A ring circuit. See Chapter 9 for ideas on circuitry.

Boiler supply

The electrical loading for the gas/oil central heating system is negligible and may be fed through a fused spur on the local socket outlet circuit. British Standards require a switch adjacent to the boiler or programmer to isolate the complete system.

Consumer unit

Residual current protection

471–16   If there are no gardens associated with flats and no likely use of portable equipment out of doors, there is no requirement for any rcd protected sockets. However, many specifiers require 30 mA rcd protection to all sockets, especially for sheltered accommodation.

314–01   If rcd protection is given to sockets it is not acceptable to use a 30 mA rcd as a main switch. This could cause nuisance tripping. Protection of sockets should be given separately either with a split bus-bar consumer unit, preferably with combined mcb/rcd units.

Circuit protection

533–01   If it is decided that no rcd protection is required, this is one situation where it may be economical for the installer to use rewirable fuses. The subject should be discussed with the client and careful note taken of the long-term main

341–01   tenance requirements of tenants. Regulations require that equipment shall be suited to the intended purpose. Rewirable fuses are not suitable for disabled or

512–06   sheltered accommodation.

Accessibility

The consumer unit must be accessible for the intended occupier. Once again note must be taken of occupier requirements.

A similar judgement should be made on the location of the cooker control

476–03   switch which may be required in an emergency. If a sink waste disposal unit is installed, this must also have an emergency switch conveniently to hand.

Earthing and bonding

Each flat has a separate electrical installation with a metered supply. An

542–04   equipotential zone must be set up within the flat. Bonding of a landlord's water and gas mains to the 100 A switch in the remote meter cupboard would not give reliable protection to the installation. It is necessary to take a full size 16 mm^2 earthing conductor to the main earthing terminal in the flat. This is shown in Fig. 3.2.

The supply company may permit the use of the third wire in a 16 mm^2 three-core steel wire armoured cable to be used as an earthing conductor on a pme supply. If a two-core cable is used, a separate 16 mm^2 earthing conductor will be required. Neither the armouring of an swa cable or the 'earth' in a 16 mm^2 twin and earth cable is large enough for the pme earthing conductor.

Main earthing terminal

The consumer's earthing terminal will be located within, or adjacent to the consumer unit. This is the point where the supply earthing conductor joins the main bonding conductors within the flat.

Bonding

Standard main bonding requirements apply for 10 mm^2 connections to
547–02 incoming water, gas or oil piping. This bonding is to be applied within 600 mm of the point of entry into the flat.
601–04   Supplementary bonding is required in the bathroom. This is best achieved in the adjacent airing cupboard, as described in Chapter 2.

PART 2 – LANDLORD'S AREAS

Meter cupboard

This cupboard is used to enclose:

❐ The electricity company's intake and metering equipment.
❐ Eleven meters and main switches for tenants' supplies.
❐ The landlord's meter and distribution board.

It is preferable for the meter cupboard to be used exclusively for electrical
513–01 equipment and it must not be used for the storage of cleaning materials. There should be clear space in front of equipment for routine meter reading and clear access for emergency attention, e.g. fire-fighting.

Supplies to flats

The loading for each flat will be handled by a 100 A switch fuse with a 63 A
523–04 BS 88 fuselink. This will protect the 16 mm^2 rising/lateral main. Cables should not be bunched and routes should avoid thermal insulation, otherwise downrating factors will require the use of 25 mm^2 cables.
    Cables should be routed in protected ducts, through public areas. It is not normally acceptable to run one consumer's cables through another consumer's property. Steel-wire-armoured cable is generally used but non-armoured sheathed cable would be acceptable provided that it is adequately protected against physical damage.

Table 3.3 Landlord's services, cable sizes and circuitry.

Circuits	mcb rating (Type 1 or 2)	Cable size (mm^2)
Lighting, lounge	6 A	1.0
Lighting, stairs	6 A	1.0
Smoke detector	6 A	1.0
Sockets		
kitchen	32 A tree	4.0
	or	
	32 A ring	2.5
lounge	20 A tree	2.5
	or	
	32 A ring	2.5
Washing machines	32 A tree	4.0
	or	
	32 A ring	2.5
Cooker	32 A	6.0
Lift	10 A	2.5

Landlord's electrical requirements

The full schedule is given in Table 3.3. Fire alarms, smoke detectors and emergency lighting are not included, these must be discussed with the relevant statutory authorities.

Diversity

313–01   The IEE Guidance Note does not show a situation that exactly fits these landlord's premises. The nearest type of accommodation listed in the Guide is that for a small hotel. Estimation of diversity is not an exact science and some judgements must be made based upon experience and consultation with the client.

It is possible that the kitchen facilities in the common room may be used fully on winter evenings, but at such times it is unlikely that the laundry equipment will be in use.

For current loading calculations, fluorescent lamp ratings must be multiplied by 1.8 to take into account control gear losses.

Lighting

Stairs, etc.	$12 \times 24\,W \times 1.8$	=	518 W
Toilets	$2 \times 24\,W \times 1.8$	=	86 W
Lounge, kitchen and laundry	$5 \times 58\,W \times 1.8$	=	522 W
Apply 75% diversity	Current	=	$\dfrac{1126 \times 75\%}{240}$
		=	3.5 A

Socket outlets

There is central heating. Kitchen equipment and a laundry iron will be used.

Assume 100% load of one circuit   =   32 A

Other equipment

Cooker at 80%   =   32 A
Washing machine and drier at 75%   =   18 A
Boiler and lift – Ignore.

Total load

Estimated maximum demand   =   Approx. 86 A

A single phase 100 A service will be appropriate.

Cable sizes and circuitry

This is a conventional building structure. PVC twin and earth 6242Y type cables are acceptable (see Table 3.3). Refer to the notes for wiring in flats for installation details, including cables in false ceiling cavities.

Lighting

The total lighting load is low. To avoid inconvenience in the event of a fault, two circuits are suggested using $1.0\,mm^2$ cables.

Socket outlets

Socket outlets will carry heavy loads in the kitchen and in the laundry. In other rooms only small appliances will be plugged in. It is suggested that three circuits will be appropriate. Table 3.3 provides for the use of both $2.5\,mm^2$ ring or $4.0\,mm^2$ tree circuitry.

Project Specification
BS 7671

Name...................... Landlord's services in block of flats

Reference................ Date.....................

---

240 V 50 Hz. TN-C-S. Supply fuse 100 A BS 1361 or BS 88

PFC less than 16 kA. Earth loop impedance less than 0.35

Consumer unit in landlord's cupboard BS 5486
100 A main switch

8 way   M6 Type 2 mcbs

Circuits	Rating (A)	Cable size (mm²)	Max run (m)	Lights/points g = gang
1. Cooker	32	6.0	43	1
2. Ring 1 kitchen)	32	2.5	71	3 x 2g
Boiler)				1
3. Ring 2 lounge & stairs	32	2.5	71	6 x 2g
4. Washing machines	32	2.5	71	2
5. Lights lounge)	6	1.0	43	7
Toilet fans)	2			
6. Lights stairs	6	1.0	43	10
7. Smoke detectors	6	1.0	43	
8. Lift	10	2.5	43	1

Figure 3.4 **Project specification for landlord's services.**

Other equipment

Separate circuits will be necessary for the cooker and the lift. Spare ways will need to be provided for emergency lighting and fire alarms. The minimal boiler supply requirement can be taken from a local socket outlet circuit.

Distribution board

A conventional domestic type is appropriate. This should be easily accessible. 513–01 If the unit is locked for security purposes, the location of the key should be clearly indicated.

Residual current protection

The Wiring Regulations only require rcd protection for sockets which may be used to supply portable equipment outdoors. To satisfy this requirement for 476–16 landlord's gardening maintenance, one integral 30 mA rcd socket could be designated for this use.

The local authority may require rcd protection for sockets in the common room which may be used for public social activities. It is also sensible to protect cleaners' sockets in public areas.

In the circumstances, the system has been designed using a split bus-bar mcb consumer unit. Circuit arrangements are shown in Fig. 3.5. Lighting circuitry and the lift supply should not be given rcd protection.

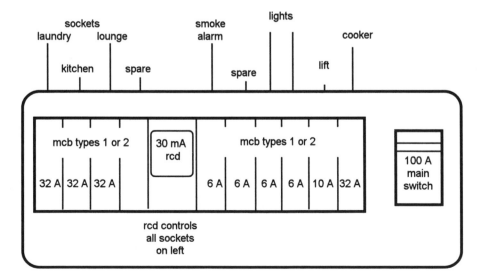

Figure 3.5   **Landlord's distribution board.**

Switchgear

Labelling

514–09 All equipment in the meter cupboard must be carefully labelled giving flat numbers corresponding to switchgear. It is probable that a 415 V label will be
514–10 required to indicate the presence of three-phase voltages between different consumers' equipment.

Switching

Standard requirements apply as in flats.

Earthing

The installation of the landlord's supply may be treated in a similar manner to
542–04 that of the supply to flats. Main bonding is required between the landlord's distribution board, main earthing terminal and other landlord's service entries, in the same way as that shown in Fig. 3.2 for flats.

Emergency systems

Details of fire alarm and emergency lighting provision should be added to the specification.

Chapter 4

Overcurrent Protection

There are two reasons for overcurrent:

Defs

- ❐ *Overload*
  This is overcurrent occurring in a circuit which is electrically sound. Examples are a user connecting too many appliances or applying excessive mechanical load to a machine.
- ❐ *Fault*
  This is overcurrent occurring as a result of a faulty installation. Examples usually involve the failure of insulation causing a short-circuit between conductors or to earth.

The two conditions are quite different by way of cause and effect. Action to lessen safety hazards is either with practical measures to detect and remedy fault conditions or to apply regulations which are intended to prevent the occurrence of overcurrent. In most circumstances, overload and fault current protection willl be given by one common device.

Overload

There are many reasons for the overloading of healthy wiring, including:

473–01
- ❐ An electric motor undertaking mechanical duty in excess of design parameters,
- ❐ Faulty running of machinery caused by bearing failure or uneven loading,
- ❐ A motor starting current,
- ❐ Loss of a phase from a three-phase load,
- ❐ Excess loading added to a socket outlet circuit,
- ❐ An underestimate of the maximum demand of an installation.

Overloads often arise gradually and at the early stages may not be apparent when testing is carried out. The transition from excess loading to overloading on a main service may result as a combination of unplanned extra sub-circuit demand.

The purpose of overload protection is to detect and clear an overloaded circuit before it becomes dangerous or damages the installation. A small overload may not cause an immediate hazard and may be tolerated by protective devices for a long time A simple example is with a 13 A fuse. This will not rupture if the load current creeps up to 13 A. It should operate indefinitely at this nominal load without producing dangerous overheating. The consequence is that it will continue to operate with a load of 18 A for some minutes. Additionally, it will tolerate even larger overload surges to avoid nuisance operation.

Thus, for example, the 13 A fuse will accept an 'instantaneous' overload of double its rating.

Overload protection

A protective device must be used to break overload current before any thermal damage is done to cable insulation or other load-carrying parts of the installation.

433–02
- The *nominal rating* of the device must be:
  Greater than the design current for the circuit and less than the operational rating of any conductor being protected.
- The *operating current* must not exceed:
  1.45 × the current rating of any conductor being protected after correction for grouping, etc..

There are some conditions where the above overload protection need not be applied to a conductor:

473–01
- Where there is a reduction in conductor size but the upstream protection gives effective protection to the smaller conductor, or
- Where the load is such that it is unlikely to be overloaded, e.g. a fixed lighting system, or
- At the intake position where the supply authority's equipment gives appropriate protection.

Overload protective devices

Rewirable fuses

Although accepted by the Wiring Regulations, preference is given to other
533–01 devices. Provided that they are correctly wired, these fuses are fail-safe and for simple situations are troublefree until abused.

One major technical disadvantage is that a cable down-rating factor of 0.725 has to be applied to a circuit protected by a rewirable fuse. Fault current breaking capacity is limited. This will be covered later in this chapter.

To avoid the above complications, rewirable fuses are not specified for projects in this book. However, if note is taken of technical and practical limitations there is no reason why they should not be used in suitable circumstances.

HBC fuses

341–01   These are fail-safe devices with excellent characteristics but liable to abuse when no spares are available. The installations covered by this book usually do not have skilled electrical maintenance personnel. For this reason HBC fusing is not specified.

HBC fuse protection will be fitted by the supply authority in sealed meter heads and HBC type g(G) fuses may be used for large main distribution fusegear as an alternative to moulded case circuit breakers.

Motor circuit protection fuses have a g(M) characteristic. This gives a
435–01   continuous operating rating and a motor starting rating. The subject is beyond the scope of this book but is an essential study for motor installers. Suppliers' technical literature.usually provides the best scource of information.

Miniature circuit-breakers

It is often said that these devices may not fail to safety, nevertheless their reliablity is beyond question for most installations. If there is no choice but to install switchgear in a damp arduous situation, HBC fuses may give better long-term operational reliability.

In all normal locations, the use of mcb distribution boards is recommended to give the most user-friendly protection.

The mcb type number

Mention has already been made of the necessity for an overload protection device to be able to withstand moderate overloading. By their nature, fuses run warm on full load and have a 'fusing factor' set by their operating character-istics. It is possible to design a circuit-breaker with an exact operational cut-off point and, in fact, a residual current device is designed to disconnect at not more than the rated current. However, this would be inconvenient for an mcb.

For large fault currents a British Standard mcb will trip *instantaneously*, i.e. within 0.1 s. There is a range of characteristics related to overload conditions. These are designated by the *type* of an mcb (see Table 4.1).

An example will indicate the method of selection:
Consider a 30 A type 2 mcb used on an office socket outlet circuit. Table 4.1 shows that this will carry between 120 A and 210 A for a few seconds before disconnection. The actual tripping current between 120 A and 210 A depends upon manufacturing tolerances and the way that the overcurrent builds up. With an overcurrent in excess of 210 A the mcb will trip in less than

Table 4.1 Type characteristics of an mcb.		
mcb type	*Tripping current range* (Factor to be multiplied by nominal rating)	*Typical application*
1	2.7 to 4	Domestic and where overload surges are not anticipated.
B	3 to 5	General purpose and commercial.
2	4 to 7	General purpose and commercial.
3	7 to 10	Motors and highly inductive discharge lighting.
C	5 to 10	Motors and highly inductive discharge lighting.
4	10 to 50	Only for special conditions, e.g. some welding plant.
D	10 to 20	Only for special conditions, e.g. some welding plant.

0.1 s. A type 2 or type B mcb is usually a suitable standard choice for most domestic and commercial installations but, as will be seen later, this depends on the earth loop impedance.

Fault current

473–02 Fault current is invariably caused by some failure in the installation. This may be because:

❑ Cable insulation has been damaged by heat or abrasion,
❑ Water has entered into a badly protected connection,
❑ A motor has burned out,
❑ A metal tool has fallen across bus-bars.

The consequences of a fault may be a fire or explosion resulting in serious burn injuries or even death.

A fault may occur between any conductors of the supply system, or to earth, or between both simultaneously. Speedy disconnection is essential.

The initial surge of fault current is usually measured in thousands of amps (kA) and is only limited by the impedance of the supply up to the fault position. This is known as the Prospective Fault Current (pfc). There will be a very short period of time before any protective device reacts to this overcurrent.

Fault current protection

The purpose of fault current protection is to disconnect the supply speedily and to restrict damage and danger as far as possible.

All parts of an installation must be protected against the highest pfc that can be anticipated at any particular point in the system. The pfc is at its highest protection at the intake position and will decrease or *attenuate* through the installation as the resistance of cables is added to the fault path.

Omission of fault current protection

473–02 In some conditions, fault current protection may be applied downstream, on the load side of part of the system. The conditions are:

❑ If the unprotected conductor up to the protection device is less than 3 m in length, and

❑ This conductor is so located that it is unlikely to experience a short circuit or earth fault, and

❑ Has superior physical protection against risks of fire and shock.

This arrangement usually applies where short and relatively small cables are connected to heavy current bus-bars to take a supply to lightly loaded fuse switches.

Enclosure in conduit or trunking is regarded as superior physical protection and the cables in question should be separated from other circuitry. Consideration must be given to the consequences of 'flash-over' between live parts.

Short circuit rating

Under short circuit conditions there will be a considerable current surge. This is the pfc referred to previously. For a standard 100 A supply the pfc may be as high as 16 000 A (16 kA). Much will depend upon the supply authority network and distance from the substation.

More usually the figure will be less than 5 kA. This current will flow until the
432–02 fuse ruptures or mcb contacts separate. During this period, before the separation arc is extinguished, the protective device must be able to withstand

Table 4.2 Breaking capacity of devices.	
Device	*Breaking Capacity* (kA)
BS 1361 fuse, Type 1	16.5
Rewirable fuses (see marking on fuse carrier)	S1 = 1 S2 = 2 S4 = 4
mcbs to BS 3871 (see marking)	M6 = 6 M9 = 9

the heat and physical stress. This is known as the breaking capacity of the device and on mcbs it is given an M rating.

Typical ratings are shown in Table 4.2.

It will be seen that rewirable fuses have a low rating. This is one of the reasons why they are excluded from the project examples given in this book.

British Standard domestic consumer units using rewirable fuses are a special case. Most units will be certified to have a conditional rating of 16 kA. This means that these units may be used for conventional housing applications, where it is known that the pfc will not exceed 16 kA. The conditions rely upon the ability of the consumer unit to withstand the stresses involved and the back-up provided by the supply authority fuse.

For simple domestic installations without the usual rcd requirement (e.g. flats), a rewirable consumer unit normally complies with the Regulations.

Disconnection times

The Wiring Regulations specify various maximum disconnection times for different types of circuits:

	Socket outlet circuits:	0.4 s
413–02	Fixed appliances:	5.0 s
	Supplies to outdoor equipment:	0.4 s
Part 6	Farm circuits:	0.2 s
	Bathroom conditions:	0.4 s

A British Standard mcb has characteristics such that it will disconnect in less than 0.1 s provided that sufficient fault current can flow.

App. 3   On the above basis, an mcb will be appropriate for any of the disconnection times covered by the Regulations.

Earth loop impedance

The prospective fault current through an earth fault depends on the maximum earth loop impedance at the point of the fault. If this is lower than figures given in the Regulations there will be 0.1 s disconnection.

413–02   Test measurements must be adjusted to take into account the increase in resistance caused by heat under fault conditions. Table 4.3 has been designed to encompass typical conditions on projects in this book. For other situations refer to the Wiring Regulations.

Summary of mcb specification

Projects covered in other chapters have selections made for each mcb to be used in the standard circumstances described.

For non-standard projects or conditions, the following data must be assessed in order to specify an mcb for overload and short-circuit protection. Similar rules apply for fuses, but for rewirable fuses an additional 0.725 de-rating factor must be applied to I_z.

❐   Establish the pfc to give the M rating
❐   Select an mcb Type to suit the application
❐   Calculate the nominal current rating I_n of the mcb as follows:

The normal load current of the circuit	$= I_b$
Use I_b to find the next highest suitable mcb	$= I_n$
I_n must be adjusted for cable grouping, thermal insulation, ambient temperature, etc.	$= I_t$
I_t will be used to select a cable with a rating	$= I_z$

❐   Establish the maximum earth-loop impedance $= Z_S$
(Adjust a 'cold' Z_s reading to take into account fault conditions.)

413–02 ❐   Check Z_s (adjusted) against I_n in the appropriate tables for Disconnection Times.

Table 4.3 Typical maximum earth loop impedance operational conditions.		
mcb BS 3871 type	Nominal rating (A)	Max. test earth loop impedance (adjusted for fault conditions) (ohms)
1	5	9
2	5	5
B	6	6
3 and C	6	3
1	15	3
2	15	1.7
B	16	2.3
3 and C	16	1.1
1	30	1.5
2	30	0.8
B	32	1.1
3 and C	32	0.6
3 and C	63	0.3

Adjustment of Z_s reading

Figures given in Wiring Regulation's tables are for test results. An adjusting factor of 0.7 takes into account temperature rise under fault conditions.

Conclusion

The choice of a cable size always follows the selection of a suitable overcurrent protection device.

In practice, rule-of-thumb cable sizes are used for conventional circuits and standardized designs as shown in this book. These cable sizes have been selected taking into account the type and rating of the overcurrent device.

Once a firm choice of cable has been made, it is not acceptable to change the mcb or fuse specification without a complete circuit redesign.

Chapter 5

An Architect's Office

This could be a local branch of a national organization or the operational headquarters of a small business. It could be the premises of an estate agent or an insurance broker.

For the purposes of this exercise the project is an architectural practice with two or three partners and a small staff. Electrical requirements are modest but include storage heaters and mains services for computers. Some ideas are given for clean lines and uninterruptible power supplies.

Lighting design will be specialized and is beyond the scope of this book. Basic lighting circuitry is given and general suggestions for extra low voltage spotlights in two areas. The layout as illustrated shows lighting and heating in order to establish loads and circuitry only. This is not a lighting or heating design scheme. Where an installer is not experienced in this type of design it is suggested that reference is made to product suppliers who usually provide a design service.

Other interested parties

Before settling the full electrical schedule a check must be made upon special or additional requirements.

❑ *Fire authority*
   Fire alarms, emergency lighting and other safety features especially if this office is part of a multi-occupancy building.
❑ *Client's insurers*
   There may be special requirements for the security of an office which contains valuable computers and documentation.
❑ *Computer specialist*
   Requirements for clean earth and uninterruptible power supply.
❑ *Landlord*
   The subject of common alarm or security systems may be of interest.
❑ *Health and safety*
   Lighting may be important where operators view computer screens for long periods.

❏ *Electricity company*

Check availability of a supply to suit the potential load and confirm the location of the intake position.

Building structure and finishes

This is a part of a new building yet to be completed. Fig. 5.1 shows the intended layout.

Total office floor area: 180 m^2.

Floor below and above: concrete beams, as yet unscreeded.

Ceiling: suspended false ceiling with 300 mm void.

Walls, external: masonry, with plaster finish; *internal*: lightweight partitions, plasterboard on steel framework.

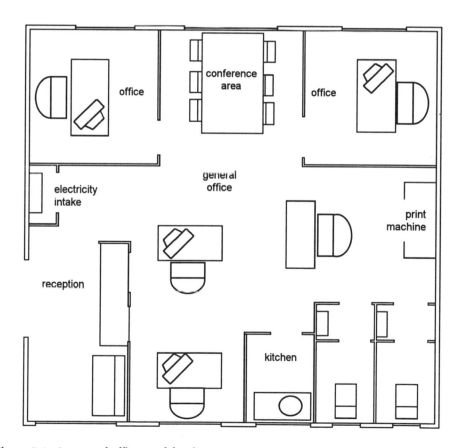

Figure 5.1 **Layout of offices and furniture.**

Table 5.1 **Electrical requirements.**	
Provisional electrical schedule	
Lighting	
General office lighting	9 × 1.8 m 70 W twin fluorescents
Two executive offices	2 × 1.8 m 70 W twin fluorescents
Conference area	9 × 50 W elv spotlights
Reception area	8 × 50 W elv spotlights
Toilets and kitchen	6 × 24 W low energy fluorescents
Direct acting heaters	
Three wall panel heaters in the kitchen and toilets, with individual thermostatic control	4 × 1 kW
Storage heaters on off-peak supply	
Nine block storage units	9 × 3 kW
Print Machine	
A single phase 20 A supply has been requested	
Socket outlets	
Some discussion and decision is required at this early stage. Two systems are suggested:	
(1) Conventional general purpose 13 A sockets. (2) Special clean lines for computers. One dedicated socket per desk place. Consideration may be given eventually to an uninterruptible power supply (UPS) service.	

Electrical requirements

A full schedule of requirements is shown in Table 5.1. Layout arrangements for lighting and power are complex. Two diagrams have been produced to supplement the schedule. Fig. 5.2 shows the preliminary lighting arrangements and Fig. 5.3 shows provisional heating and power arrangements. The requirement for off-peak storage heaters is noted and a suggestion of clean lines for computers. It is the responsibility of the occupier or developer to consult with the supply authority to be certain that an appropriate electricity supply will be available. This task is normally delegated to the electrical

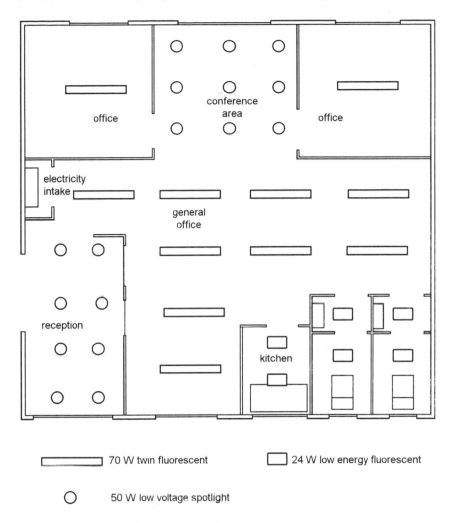

70 W twin fluorescent   24 W low energy fluorescent

50 W low voltage spotlight

Figure 5.2   **Provisional lighting layout for electrical design purposes.**

contractor who will need to establish supply requirements including the anticipated maximum demand. A Project Specification format is shown in Fig. 5.4. This, together with layout drawings, becomes part of the contract document.

Loading and diversity

This is not an exact science and calculations may be rounded off. Every project must be considered taking into account working conditions.

311–01   A small office situation illustrates the condition where lighting and heating loads are likely to be used simultaneously at maximum capacity on certain

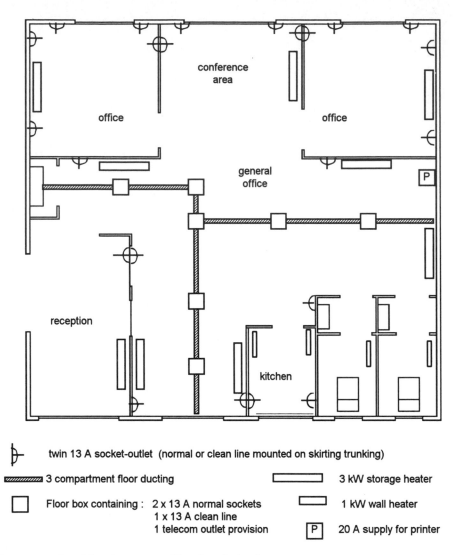

Figure 5.3 **Provisional power and heating design layout.**

Project Specification BS 7671 Office Unit

Name....................... Location.......................   Reference................. Date.....................

Three phase 240 V 50 Hz. TN-C-S.   Supply fuse 100 A BS 1361 or BS 88

PFC less than 16 kA. Earth loop impedance less than 0.35   100 A TP main switch   Two Type B distribution boards   Normal and off-peak supplies

Normal supply

Circuits	Rating (A)	Cable size (mm^2)	Max length (m)	Lights/points g = gang
1. Sockets 1	32	2.5	71	10 x 2g
2. Sockets 2	32	2.5	71	8 x 2g
3. Sockets floor	32	2.5	71	7 x 2g
4. Sockets clean line	32	2.5	71	5 x 2g
5. Lights 1	10	1.5	55	11
6. Lights 2	10	1.5	55	13
7. Lights 3	10	1.5	55	10
8. Print machine	20	2.5	45	
9. Heaters in toilets	20	2.5	45	4

Off-peak service

1-9 Storage heaters	20 A	2.5	25	9

Figure 5.4 **Project specification for offices.**

occasions. The off-peak heaters may also be brought on line at a time when the office is working. No diversity can be allowed on these loads. However, it is probably acceptable to ignore the lighting and thermostatically controlled heaters in toilets.

Lighting

For current loading calculations, fluorescent lamp ratings must be multiplied by 1.8 to take into account control gear losses.

$$\text{Fluorescents} \quad \frac{22 \text{ lamps} \times 70 \text{ W} \times 1.8}{240} = 11.5 \text{ A}$$

$$\text{ELV spotlights} \quad \frac{17 \times 50 \text{ W}}{240} = 3.5 \text{ A}$$

Storage heaters

$$\frac{9 \times 3000 \text{ W}}{240} = 112.5 \text{ A}$$

Print machine

$$20 \text{ A supply} = 20 \text{ A}$$

Socket outlets

These will only be used for desktop equipment. Any reference to floor area or number of outlets is meaningless. A suitable estimate of desktop loading is between 1 A and 3 A per station. Some diversity can be written in.

Assume here 2 A per desk, with room for eight desks   =   16 A

Total load

Total current (approx.)   =   164 A

This will be distributed across three phases and a three-phase 100 A supply will be appropriate. It is understood that the electricity company will give a pme service.

Wiring system

Discussions with the client indicate that the best method of socket outlet distribution would be with electrical skirting and underfloor ducting with outlet boxes. To a large extent the building construction has determined the wiring systems to be used.

Skirting system

528–01 This will be a proprietary steel or non-metallic system. A rigid PVC skirting may be most attractive for a prestigious office. Three compartments are necessary throughout with suitable segregated door-crossing adaptations.

Compartments will be used for:

- ❑ General power circuitry and storage heaters,
- ❑ The computer clean line,
- ❑ Telecommunication and data cables.

Provided that the skirting is installed to give a completely fire-resistant enclosure, ordinary unsheathed 6491X PVC-insulated cables may be used. Special care is needed at corners and junctions to ensure that the cable enclosure is complete. Socket outlets will be mounted directly on the skirting at positions related to user requirements. The skirting design should be suitable for future outlet additions.

Underfloor system

521–07 This will be a three compartment system with segregated floor access boxes. These floor boxes have been planned on a 3 m module. Office users will either find socket outlets specially located for desk positions or within 1.5 m.

The mains cables will be fully protected and enclosed within floor trunking. Standard unsheathed 6491X, PVC-insulated cables may be used.

Socket outlets

Two systems will be in use:

- ❑ The general-purpose 13 A standard BS 1363 type, and
- ❑ A special 13 A plug with non-standard pin orientation for the dedicated 'clean' service to computers.

It is recommended that the wiring for this clean line should, wherever possible, take a different route or be contained in a segregated trunking compartment. A dedicated circuit protective conductor will be taken to isolated earthing pins on the special sockets. This conductor will be insulated green/yellow throughout its length and finally connected to the main earthing terminal for the installation.

Lighting circuits

There are various wiring possibilities. A choice of two is given below. In both
553–04   cases the use of luminaire support connectors (LSCs) is recommended. These
provide a good accessible plug-in facility for maintenance work.

(1) Conduit and trunking system

Outlet positions will be fixed to the underside of the soffit and interconnected
with plastic trunking or plastic conduit fixed to the ceiling. A flex connection is
then made from these ceiling outlets to luminaires or extra low-voltage
transformers.
Plastic conduit drops will be used for services to wall switches.

(2) Sheathed twin and earth cables

Twin and earth cable can be used, concealed in the ceiling void. It must be
522–08   fixed to the ceiling either directly or on a timber batten and not draped across
a metallic ceiling grid.
The cable can be taken straight into the luminaire connector or transformer
526–03   terminals provided that there is no excessive heat and the cable is protected
against abrasion at the entry point. This tends to be an untidy installation and
130–01   note must be taken of the Wiring Regulations requirement for good workmanship.
A much tidier job can be achieved with the use of ceiling roses or luminaire
supporting couplers with flex connections to luminaires and transformers.

Battened out ceilings

Where headroom is at a premium, or for reasons of economy, false ceilings
are often constructed of plasterboard fitted to timber battens. Wiring run in
522–06   this space is subject to restrictions similar to those imposed on wiring installed
beneath floor boarding.

(1)   Unprotected cable must run at least 50 mm from the underside of the
batten, or
(2)   Cables must incorporate an earthed sheath (mics or swa), or
(3)   Cables must be enclosed in earthed steel conduit or given equivalent
mechanical protection.

Options 2 and 3 are expensive and may be impracticable. In order to use
unprotected twin and earth cable a convenient method for lowering the
ceiling to comply with the 50 mm requirement is to cross-batten the ceiling. If
battens are orientated in the right direction, this has the added advantage that
cables may be easily routed without the need to drill or notch-out wiring
channels.

Low voltage lighting

Two office areas have been planned to have decorative extra low voltage lighting.

This scheme is intended to give guidance upon wiring practice. The actual number and selection of luminaires will be a special design detail and is not included here. References should be made to suppliers regarding SELV lighting systems.

There is a choice of lv transformer systems.

Group transformers

This involves the use of relatively large transformers, each of which will supply a group of luminaires. This may have cost advantages but rating is important.

  ❏ LV lamps are voltage critical. Only the designated number of luminaires may be connected to the transformer and with some designs the output voltage will rise upon failure of one lamp. This will shorten the life of other lamps connected to the same transformer.

  ❏ The transformer may run hot and good ventilation will be essential. There
422–01 must be adequate clearance above and around the unit. Manufacturers' instructions should give this information.

526–04 ❏ There must be access to transformer terminations. The length of cable runs is critical and this may cause problems in locating a good accessible mounting position.

  ❏ Wiring between the transformer and a luminaire must be completely
413–06 segregated from other cables. The connection to the fitting must be contained in a fire-resistant enclosure and the final tails to the lampholder
528–01 will need to be heat resistant.

526–02 ❏ The wiring system is subject to SELV regulations which include no pro-
411–02 vision for earthing.

Individual transformers

These are matched to single luminaires; they usually run cool and have protective thermistors. Probably the most convenient is the type that fits through the luminaire mounting aperture and stands alongside within the ceiling cavity.

Fire prevention

All flush-mounted luminaires which project to the rear of ceiling linings must be enclosed and have adequate ventilation. Care must be taken to ensure that no combustible materials can come into contact with hot surfaces and that thermal insulation does not restrict ventilation.

Table 5.2 Arrangement of circuits across phases.		
Phase		*Amps* (approx.)
Red	*Lights*	
	Executive offices and conference area	4.0
	Storage heaters	37.5
	Print machine	20
	Possible total	62
Yellow	*Lights*	
	Centre office and toilets	8.4
	Storage heaters	37.5
	Window wall sockets	10
	Inside wall sockets	10
	Possible total	66
Blue	*Lights*	
	Inside office and reception	3.7
	Storage heaters	37.5
	Floor sockets	10
	Dedicated sockets (clean line)	10
	Heaters in toilets	12
	Possible total	71

Arrangement of circuits

Table 5.2 indicates a possible spread of circuits across phases. There is flexibility in the arrangement but it is sufficiently accurate for general applications. Actual conditions will vary with inevitable load diversification.

Distribution boards

There will be two, Type B, three phase distribution boards to accommodate normal and off-peak services (see Fig. 5.5). These will be located at the meter position. The office layout is compact and no sub-mains are required. None of the equipment has any especially high starting loads, therefore type 2 mcbs will be appropriate. There is no requirement for rcd protection and the use of an rcd could cause nuisance tripping, especially on the computer service.

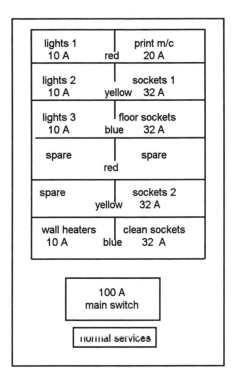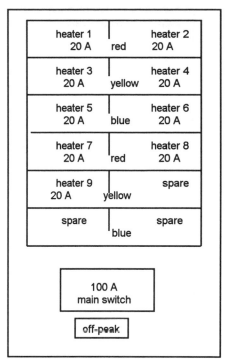

lights 1 10 A	red	print m/c 20 A		
lights 2 10 A	yellow	sockets 1 32 A		
lights 3 10 A	blue	floor sockets 32 A		
spare	red	spare		
spare	yellow	sockets 2 32 A		
wall heaters 10 A	blue	clean sockets 32 A		

100 A main switch

normal services

heater 1 20 A	red	heater 2 20 A		
heater 3 20 A	yellow	heater 4 20 A		
heater 5 20 A	blue	heater 6 20 A		
heater 7 20 A	red	heater 8 20 A		
heater 9 20 A	yellow	spare		
spare	blue	spare		

100 A main switch

off-peak

Figure 5.5   **Two distribution boards for office.**

Cable sizes

Cable sizes are given in Table 5.3. It will be seen that these relate to:

App. 4

- ❏ The maximum design current for the circuit or appliance
- ❏ The rating of the protective mcb
- ❏ Installation methods and enclosures
- ❏ Grouping as shown
- ❏ Maximum length of run
- ❏ Maximum earth loop impedance of the pme supply at 0.35 ohms
- ❏ Voltage drop.

If any of these variables change or exceed the limits shown, sizes must be re-calculated.

Switchgear

Ch. 46   Standard requirements for isolation and switching apply.

Table 5.3 **Cable sizes.**

Circuit	Full load (A)	mcb rating (A)	Cable type (ref.)	Size (mm²)	Max. length (m)
Lighting					
(Enclosed in conduit or trunking)					
1	4.0	10	singles 6491X	1.5	55
2	8.4	10	singles 6491X	1.5	55
3	3.7	10	singles 6491X	1.5	55
(Unenclosed)					
1	4.0	10	T & E 6242Y	1.5	55
2	8.4	10	T & E 6242Y	1.5	55
3	3.7	10	T & E 6242Y	1.5	55
Power sockets Floor and wall outlets Alternative possibilities with cables enclosed in skirting or underfloor trunking					
3 circuits each	20 A max.	20	Radial singles 6491X	2.5	30
3 circuits each	32 A max.	32	Radial singles 6491X	4.0	45
3 circuits each	32 A max.	32	Ring singles 6491X	2.5	70
Dedicated power circuit for computers As above for power circuits with cables enclosed in skirting or underfloor trunking					
Storage heaters All similar with a maximum of three circuits enclosed in skirting trunking					
9 × 3 kW	12.5	16	singles 6491X	2.5	45
Wall heaters One circuit, three heaters with cables enclosed in skirting trunking					
3 × 1 kW	12.5	16	singles 6491X	2.5	45

Print machine

A 20 A isolator makes provision for this machine at the point of connection.

Wall heaters in toilets

All three 1 kW heaters have individual thermostats and will be fed off the same circuit. They must be fused locally. The most convenient arrangement will be
432–02 to have a switched fused connection unit adjacent to each heater.

Storage heaters

There is a separate circuit for each storage heater. No local fusing is required and if the heater has user controls, no additional functional switching is required, or switching for mechanical maintenance. Isolation for electrical
476–02 work can be carried out by securing the appropriate circuit-breaker in the off position.

Presence of 415 volts

Although it is good practice to separate items of single-phase equipment and accessories which have been connected to different phases, the Regulations do not prohibit such connection.

Where separate items of equipment have between them voltages in excess
514–10 of 250 V at terminals that are simultaneously accessible, a suitable 415 V warning label must be applied. The label is not of consumer interest but is to warn any person having access to live parts. The most secure place for a label is under the terminal cover or within a switchbox.

Large items of three-phase switchgear should have external 415 V labels.

Access to switchgear

In this instance the main switchgear is located in a cupboard. Unless there is
513–01 no possibility of confusion, the cupboard door should be labelled to indicate the presence of electrical equipment. Normally the door should not be lock-
514–08 able, but where this is found to be necessary, a clear indication of where the key may be found should be given for emergency access.
130–07 Wherever possible the occupants should be advised of the need to keep clear access to controls.

Earthing and bonding

This unit is part of a multi-occupancy building complex, nevertheless this individual electrical installation must have the full earthing treatment. It may well be that water supplies are fed from a main riser which is bonded to the landlord's electrical service elsewhere, but regardless of this, the water pipe must be bonded at the point of entry into the architect's office.

Main earthing terminal

A single main earthing terminal is required which will be used to service both normal and off-peak installations. It should be possible to disconnect the earth from either installation without intefering with the other; therefore the terminal must be a separate item and be accessible for test purposes without exposing live parts (see Fig. 5.6).

542–04

A 16 mm^2 main earthing conductor from the supply pme earth will be taken to this main earthing terminal. Main bonding conductors at 10 mm^2 are required from the main earthing terminal to:

547–02

- Mainswater stop cock,
- Main gas stop cock (if applicable).

Additionally

543–01

- Two 10 mm^2 (minimum) circuit protective conductors will connect to the earthing bars in the distribution boards,

- A special 2.5 mm^2 earthing conductor will be provided for the clean earth on dedicated socket outlets.

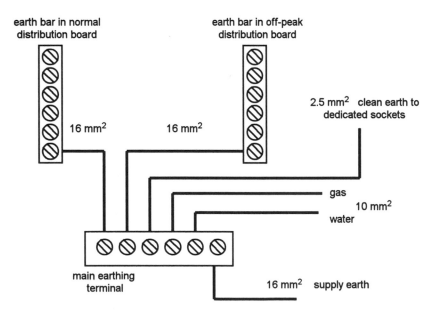

Figure 5.6 **Earthing arrangements for office.**

False ceiling grid

Previous mention has been made of the protection of wiring within a plasterboard false ceiling. Alternatively there may be a proprietary metallic ceiling grid with inset tiles. This raises the issue of bonding and the definition of an extraneous conductive part.

The definition of an extraneous part implies that it is liable to bring a different earth potential into an otherwise equipotential zone. This would not apply to a metallic ceiling grid which is entirely within the office equipotential zone. There is no requirement for main or supplementary bonding.

If the ceiling grid is carrying lighting fittings, they will have circuit protective conductors which may or may not make fortuitous contact with the grid. In either case this is of no consequence as far as the equipotential bonding is concerned. If there are no lights in contact with the grid, it will be isolated metalwork within the zone and no bonding is required.

547–02   If it is the case that a ceiling grid is continuous into the landlord's corridor or another equipotential zone, it will be necessary to bond the office grid to the main earthing terminal for the office to ensure that it could not carry an imported potential into the office equipotential zone.

Clean supply for computers

A clean supply requires a separate *functional earth* which does not carry *protective earth* currents. This conductor will be taken from the main

542–04   earthing terminal, directly to the isolated special earth terminal on dedicated non-standard sockets (see Fig. 5.6).

607–05   Any requirement for protective earthing on exposed conductive parts of equipment should be connected through the normal earthing facility. This will

413–02   apply to the print machine and photocopier.

The only point of interconnection between the protective earth and the functional clean earth should be at the main earthing terminal.

Computer installations

Apart from giving a clean earth connection, there are other implications to the use of Information Technology equipment in offices. These apply to a

Sec. 607   greater or lesser extent according to whether a particular office computer is a basic word processor or if it is part of a 'networked' system involving data exchange with other equipment on or off the premises.

High earth leakage currents

In order to filter out mains spike and surges, computerised equipment incorporates capacitor bridge circuitry connected to earth. This means that each item will have an inbuilt and normally harmless earth leakage current in the order of a few milliamps.

This subject is dealt with in depth in Chapter 7 and should be noted for large office installations. On this small project, the few desk-top computers will produce minimal earth leakage which will be handled by the specified earthing arrangements.

Mains filters

512–05   Special provision may be required for filtered mains supplies to desktop computers, especially where mains-borne interference is significant. Spikes and transients from external supply network loads can cause problems with both hardware and software.

Filtered, switch socket outlets are available to deal with electronic noise. These could be fitted at all points where computers may be plugged in. Alternatively low cost special plug adaptors are available.

Uninterruptible power supplies (UPS)

Unexpected power failure causes loss of computer data. The installation of 313–02   basic UPS equipment will supply power for a limited period of time to enable the operator to carry out regulated shut-down procedures. An operational period of five minutes may be adequate in normal circumstances. More expensive equipment will give continuity of supply for longer periods.

The client may choose to have centralised UPS equipment with all dedicated sockets connected to a system which can carry the full load. Alternatively, single desktop units may have a local individual UPS source.

The subject should be discussed at the quotation stage of the project to ensure that mains switchgear and wiring takes account of UPS distribution requirements. It is always preferable to segregate UPS cables from normal 331–01   power cables and communications circuitry. The topic is becoming increasingly important with new applications for Information Technology equipment.

Chapter 6

A High Street Shop

A small shop with just a counter and storage areas is simple to design. The electrical system follows the layout for a house but care must be taken on loading factors if direct acting electric heaters are switched on all day.

These days, many shops have specialized equipment for cooking food or providing other services such as shoe repairs or photographic processing. It is this type of shop that sometimes causes problems for the electrical installer in the estimation of loads.

This exercise takes a High Street bakery shop as a typical project. The design may be adapted for similar conditions.

Special considerations

The electrical contractor is often given a shop-fitter's layout plan with minimal electrical information. This is not sufficient to even offer a quotation.

For design purposes it is essential that the electrical contractor has a full summary of services required. There may be specialized lighting and a mix of single and three-phase machinery. The fact that food is processed on the premises implies strict hygiene rules which may affect surface runs of conduit. The public have access to the shop, therefore equipment must be suitably located away from unauthorized interference.

Early consultation with the client is necessary. Sometimes equipment is brought from old premises and most clients should have some idea of their actual requirements.

Other interested parties

Before settling the full electrical schedule a check must be made upon special or additional requirements.

❑ *Local authority*
  Food preparation hygiene facilities.
❑ *Fire authority*
  Fire alarms, emergency lighting and other safety features.

❏   *Client's insurers*
   Safety equipment in the shop and food preparation areas.
❏   *Landlord*
   Common alarm or security systems may be used in a shopping complex.
❏   *Health and safety*
   There may be safety restrictions regarding the use of machinery in some
   areas.
❏   *Electricity company*
   Check availability of a supply to suit the potential load and confirm the
   location of the intake position.

Building structure and finishes

❏   Total floor area, 100 m^2.
❏   Concrete floor, tiled throughout.
❏   Concrete soffit, exposed in bakery; suspended false ceiling in shop area.
❏   Walls, brick or building block. Fair-faced exposed in bakery; tiled or
   plastered in shop and toilets.
❏   Space heating by gas.

Electrical requirements

Proposed electrical layouts are shown in Figs 6.1 and 6.2. A full schedule of
requirements is shown in Table 6.1. It will be noted that space heating is by
gas although cooking is electric. This may not be realistic. In practice a more
likely situation would have gas cooking and heating. However, many catering
establishments do have mixed services and in this case the electric cooking
has been chosen in order to calculate diversity. A suitable Project Specification is shown in Fig. 6.3.

Loading and diversity

It will be seen that the assessment of maximum demand is very much a matter
311–01   of experience. There is a tendency to overestimate high fixed loads which in
practice only occur for short periods of time. Thermostats and energy
regulators switch heater elements on and off at irregular intervals.
   The biggest load on this project is the oven in the kitchen and it is extremely
unlikely that the fully loaded 10 kW condition will coincide with full loading on
other appliances.

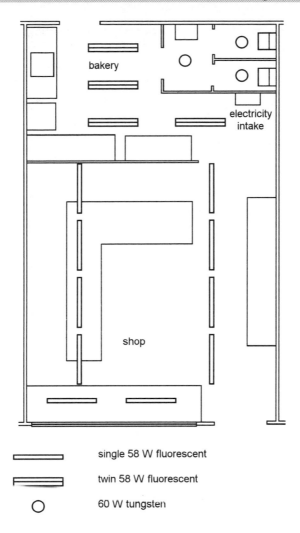

bakery

electricity
intake

shop

single 58 W fluorescent

twin 58 W fluorescent

60 W tungsten

Figure 6.1 **Lighting layout for wiring purposes only.**

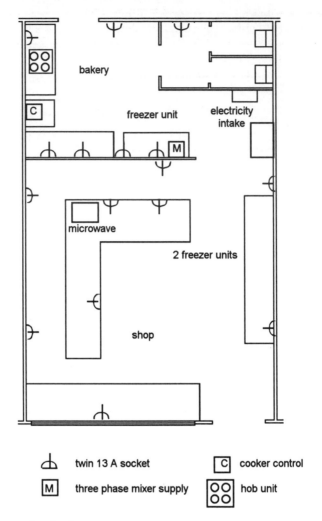

bakery

freezer unit

electricity
intake

C

microwave

2 freezer units

shop

| | twin 13 A socket | | cooker control |
| M | three phase mixer supply | | hob unit |

Figure 6.2   **Power outlets in shop.**

Table 6.1 Provisional electrical schedule.	
Shop	*Bakery*
Lighting	
10 × 58 W	8 × 58 W 4 × 60 W
Twin 13 A sockets	
Display units 2 Microwave oven 1 Freezers 2 General purpose 6	Freezer 1 Refrigerator 1 Small mixers 2 General purpose 4
Other loads	
	Three-phase mixer, 750 kW Oven, 10 kW Hob unit, 4 × 1.5 kW

Information is required to determine the size of mains supply. If there is any doubt it is worth showing diversity figures to the supply company.

The IEE Selection and Erection Guidance Note gives some advice about loading diversity. The figures used here have been taken from the Guide, but every project must be considered separately to take into account special factors. This is not an accurate exercise and all figures have been rounded off.

This installation will of necessity be a three-phase load and maximum current per phase is the important calculation.

Lighting

An allowance of 90% diversity is acceptable. For current loading calculations fluorescent lamp ratings must be multiplied by 1.8 to take into account control gear losses.

An assumption of 100 W per outlet is made for tungsten lamps regardless of the specification lamp size. For convenience the bakery shop lighting is spread over two circuits.

	Project Specification BS 7671 High Street Shop

Name...................... Location.......................

Reference................ Date......................

Three phase 240 V 50 Hz. TN-C-S.
Supply fuse 100 A BS 1361 or BS 88

PFC less than 16 kA. Earth loop impedance less than 0.35

100 A TP main switch

Type B distribution board

Circuits	Rating (A)	Cable size (mm^2)	Max length (m)	Lights/points g = gang
1. Sockets shop 1	32	2.5	71	4 x 2g
2. Sockets shop 2	32	2.5	71	5 x 2g
3. Sockets bakery	32	2.5	71	7 x 2g
4. Lights 1	10	1.5	55	11
5. Lights 2	6	1.0	55	13
6. Lights 3	6	1.0	55	10
7. Hob unit	32	6.0	40	
8. Oven	50	10.0	40	
9. Mixer 3-phase	6	1.0	55	

Figure 6.3 **Project specification for shop.**

Shop lighting load

$$= 10 \times 58\,\text{W} \times 1.8$$
$$= 1044\,\text{W}$$

at 90% diversity $\dfrac{1044 \times 90\%}{240} = 3.9\,\text{A}$

Bakery lighting

$$= (8 \times 58\,\text{W} \times 1.8) + (3 \times 100)$$
$$= 1135\,\text{W}$$

at 90% diversity $\dfrac{1135 \times 90\%}{240} = 4.3\,\text{A}$

Socket outlets

Three circuits would be appropriate. It is customary to put all sockets on one phase. There is no requirement for this in the Regulations but at this stage phase balancing has yet to be considered. Diversity allowances are 100% of first circuit and 40% of all others.

$$30 + 12 + 12\,\text{A} = 54\,\text{A}$$
$$= 18\,\text{A per circuit}$$

Note that this figure is for total load calculations only, not circuit cable sizing.

Other appliances

Mixer, at 50% diversity

$$\frac{750\,\text{W} \times 50\%}{240 \times 3} = 0.5\,\text{A/phase}$$

Oven, at 100% diversity

$$\frac{10\,\text{kW} \times 1000}{240} = 41.6\,\text{A}$$

Hob unit, at 80% diversity

$$\frac{6\,\text{kW} \times 1000 \times 80\%}{240} = 20\,\text{A}$$

Phase balance

It is essential to balance loads across three phases as far as possible. This exercise should consider the diversified current demands to obtain a balance under normal working conditions.

Table 6.2 Arrangement of loads taking diversity into account.		
Phase		Amps
Red	Sockets	54
	Mixer	0.5
		54.5
Yellow	Lights	8.2
	Hob	20
	Mixer	0.5
		28.7
Blue	Oven	41.6
	Mixer	0.5
		42.1

Table 6.2 gives an arrangement which in the circumstances is reasonable but not a good balance. Note that at this stage, this does not necessarily indicate final distribution board particulars. It may be better to put ring circuits on different phases.

Wiring systems

This project is interesting in that, at this early stage, some consideration must be given to the type of wiring systems appropriate to the two main areas, shop and bakery. This is one subject upon which the landlord or insurer may have an interest.

Start by considering cost

From the financial viewpoint it is sensible to consider the most economical wiring system that the Regulations will permit. This is the starting point for design. Changes to more sophisticated methods may be introduced as the situation dictates.

Undoubtedly twin and earth cable is the easiest and cheapest system to install, but the outer sheathing gives only limited mechanical protection. In the case of the shop, a certain amount of physical maltreatment should be anticipated.

522–06

Shop area

- ☐ The public have access and must not be put at risk by their own activities.
- ☐ Children will be present and inquisitive fingers can cause damage.
- ☐ The shop staff may be heavy-handed in cleaning or careless when handling trays of stock.
- ☐ This food shop is subject to hygiene standards which may include frequent washing down.

516–06 Surface mounted twin and earth cable is unsuitable for this situation. It would be precluded by regulations which specify that the electrical system shall be appropriate to the situation and the method of installation shall take into account the conditions likely to be encountered.

It is acceptable to install unexposed twin and earth cables under plaster or within building voids. This applies on the shop side of this project. Here the concealed cables are deemed to be protected by their location.

Bakery area

Different parts of the bakery area are subject to differing conditions.

- ☐ Some general physical abuse can be expected.
- ☐ High temperatures will occur in close proximity to the oven and hob units.
- ☐ Wall surfaces may be frequently washed down

In these circumstances, the choice of surface wiring systems is one or more of:

- ☐ Steel conduit and trunking,
- ☐ Plastic conduit and trunking,
- ☐ Mineral insulated, copper-sheathed cable (mics) with pvc outersheath,
- ☐ Steel wire armoured PVC cable.

Temperature limit of 70°C

523–01 A point of interest is that wherever general purpose PVC cable is used, the recommended limiting temperature is 70°C. This will apply to all the systems shown above including the mics with PVC outer sheath. Therefore none has any working temperature advantage.

423–01 A surface heated to 70°C is very hot and would only exist on an oven casing. The Regulations show 70°C as the limiting temperature for accessible parts of metallic enclosures for electrical equipment. An air temperature in excess of 70°C will only be found in the oven or above a hob unit. There would appear to be no problem of cable selection on the basis of temperature alone.

Temperature limit of 90°C

If a higher temperature rating is considered to be important, the use of 90°C XLPE cable is possible. This insulation is available on single-core conduit cable. Care should be taken with armoured cable with XLPE insulation. Off-the-shelf supplies will probably have PVC sheathing.

In theory, rigid PVC conduit and trunking could also be used at 90°C but the fixing arrangements to accommodate expansion and prevent sagging would be impracticable.

Final selection and cable sizes

Having taken all factors into account, the final selection of wiring systems on this project can be made. This is shown in Table 6.3. It should be noted that cable sizes and maximum lengths are based upon the scheme as illustrated:

❑   0.35 ohms maximum earth loop impedance,
❑   Maximum voltage drop of 4%,
525–01 ❑   Type 2 mcbs.

Bakery wiring

❑   *Cable*
     70°C PVC-insulated Ref. 6491X, single-core cables.
❑   *Steel or PVC trunking*
522–02   Main distribution from distribution board at high level to avoid any heat from appliances.
❑   *Heavy gauge PVC conduit*

(1)   Drops to outlets at worktop mounting height,
(2)   Ceiling mounted exposed to luminaires.

As an alternative, and to make cleaning easier, PVC mini-trunking could be used for wall drops.

Shop wiring

❑   *Cable*
     Twin and earth PVC-insulated Ref. 6242Y.

(1)   Cables clipped to soffit within false ceiling,
(2)   Plaster depth down walls with PVC or steel capping.

Table 6.3 **Proposed cable sizes.**					
Circuit	*Full load* (A)	*mcb* (A)	*Type ref*	*Size* (mm)	*Max length* (m)
Bakery lights in conduit and trunking.	4.7	6	Singles 6491X or Singles 6491X	1.0 1.5	36 55
Shop lights, surface or embedded in plaster.	4.3	6	T & E 6242Y or T & E 6242Y	1.0 1.5	36 55
Baker sockets, ring in conduit and trunking.	—	32	Singles 6491X	2.5	66
Shop sockets, ring, surface or embedded in plaster.	—	32	T & E 6242Y	2.5	66
Mixer 3-phase, in conduit and trunking.	3.2	6	Singles 6491X	1.0	100
Hob, in conduit and trunking.	25	32	Singles 6491X	6.0	40
Oven, in conduit and trunking.	42	50	Singles 6491X	10.0	40

Distribution board

This is a small layout and no sub-mains are required. Switchgear selection must take account of the total connected load and not take diversity into account. A suitable distribution board is shown in Fig. 6.4.

There is at least one three-phase appliance; therefore a type B board is appropriate.None of the equipment has any especially high starting loads. Standard miniature circuit-breakers, type 2, are suitable.

471–16 If there is no reasonable possibility of portable equipment being used out of doors, no rcd protected sockets are necessary. If however outside equipment

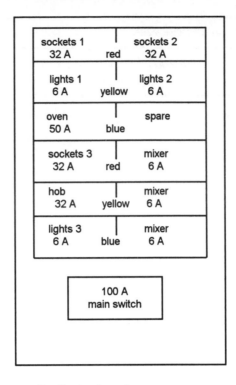

Figure 6.4   **Arrangements at distribution board.**

may possibly be used, the easiest method of compliance would be with an rcd protected socket adjacent to the back door.

Cable sizes

App. 4
525–01
Sizing must take account of mcb rating, any applicable derating factors, earth loop impedance and voltage drop on full load. Suitable cable sizes are shown in Table 6.3. Maximum lengths of run should be noted if the design is used for a different layout.

Switchgear

512–06
Care should be taken in the location of the metering and main distribution board. Conditions may be dusty with steam on occasions. The installer should recommend a cupboard to enclose the switchgear.

Isolation and switching

Main switch

460–01 This should be three pole. Switching of the neutral is not required.

Appliances

Cooker, hob unit and power mixer

All three appliances require individual treatment:

461–01 ❐ Switching for isolation,
462–01 ❐ Switching for mechanical maintenance,
463–01 ❐ Emergency switching.

All of these functions can be carried out with a suitably rated local isolator. For Sec. 537 the emergency switching function this isolator must be immediately accessible to the user of the appliance. It is quite possible that the mixer will be supplied through a three-phase BS 4343 plug and socket. This connection would be acceptable as the means of disconnection for isolation and mechanical maintenance, but not for emergency switching.

The mixer has a rating in excess of 0.37 kW and is required to have no-volt provision. This may be incorporated in the machine or with an external starter. The starter stop button would then be acceptable as the emergency stop device.

Deep freezers, refrigerators and microwave ovens

Plug and socket disconnection is adequate for all of these appliances. To avoid accidental disconnection it may be desirable to plug deep freeze cabinets into unswitched socket outlets, or use fused connection units.

Earthing and bonding

547–02 Standard arrangements apply for a pme service.

Main earthing terminal

542–04 The 16 mm^2 main earthing conductor from the supply earth is taken to the main earthing terminal (MET) (see Fig. 6.5).

Main bonding conductors at 10 mm^2 are required from the MET to:

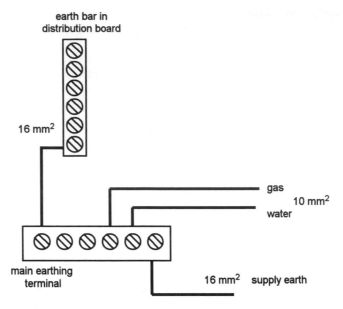

Figure 6.5   **Earthing arrangements for shop. (The two bars may be combined if accessible.)**

- ❏   Main water stop cock,
- ❏   Main gas stop cock,
- ❏   Structural steelwork (if any).

False-ceiling grid

The Wiring Regulations' definition of extraneous conductive parts would not apply to a metallic false ceiling grid which is entirely within the existing equipotential zone. There is no requirement for main or supplementary bonding. This subject is explained in detail in Chapter 7.

Steel tables in the bakery

Using the same logic as for ceiling grids free-standing, stainless steel, tables do not require bonding – even if electrical appliances are used. It is possible that a table bolted to a concrete floor could import an earth potential. In such a case bonding may be necessary.

If it is considered to be advisable to cross bond metal tables, the sizes of protective conductors should be shown in the contract documentation.

Chapter 7
Earthing and Bonding

The subject of earthing and bonding is complex and raises many controversial issues. The Bill Atkinson rules give only one interpretation of the 16th edition of the Wiring Regulations. These are offered in good faith as a starting point for specifications and contract negotiations.

IEE Guidance Note No.5, *Protection Against Electric Shock*, gives a general interpretation of the Regulations and is a good reference document, but there is no definitive interpretation for *every* situation. A competent person must be prepared to make a judgement which ultimately may be tested in a court of law.

Customers and specifying engineers are entitled to state special requirements for contracts. Their requirements are often in excess of the Wiring Regulations and there is no harm in this provided that contract documents show what is actually required, and the customer pays for the work. If a project specification only calls for compliance with the Regulations, the contractor may make a judgement as to the minimum requirements. This does not imply a low standard. The Wiring Regulations set a high standard with built-in safety factors.

Ch. 4

This chapter illustrates the use of these minimum requirements by explaining the reasoning behind earthing and bonding principles. These principles form the first line of electrical shock protection. They should be studied carefully.

Take no chances. If in doubt upon earthing requirements, seek advice.

Terminology

Explanations in this chapter stand alone. They are based on the Wiring Regulations but may refer to more than one Regulation Number.

It is important to use the correct Wiring Regulations terminology for earthing, especially if this is different from personally familiar words and phrases. Disputes are sometimes caused by a misunderstanding of IEE definitions. Quite often the result is that participants talk at cross purposes. Confusion is even more significant where the wrong words are used in written documentation, especially when these are contract requirements.

Unfortunately the definitions of many words have changed in order to give internationally harmonized meanings. We must all learn the new language. We may be convinced that our own long-established usage is better than the official interpretation, but a court or tribunal will not settle a claim on personal whims.

Definitions

This is one occasion when it may help to give some actual definitions from the Wiring Regulations:

☐ *Earthing*
The act of connecting exposed conductive parts of an installation to the main earthing terminal of the installation.

☐ *Earthing conductor*
A protective conductor connecting the main earthing terminal of an installation to an earth electrode or to other means of earthing.

☐ *Equipotential bonding*
Electrical connection maintaining various exposed conductive parts and extraneous conductive parts at substantially the same potential.

☐ *Bonding conductor*
A protective conductor providing equipotential bonding.

Green and yellow conductors

The most important conclusion to be made from the definitions is that the words *earthing* and *bonding* are not synonymous or interchange-able. The correct group term for green and yellow cables is *protective conductors*.

514–03

A protective conductor may be:

☐ The *main earthing conductor* for the installation. This connects the consumer's main earthing terminal (met) with the supply earth.
☐ A *main bonding conductor*. This connects the met to extraneous con-ductive parts, e.g. water and gas mains.
☐ *Supplementary bonding*. This is not normally required except in special locations where there is increased shock risk.
☐ *Circuit protective conductors*. These are the connections to exposed conductive parts and include the metallic parts of an electrical system and appliances frames.

Equipotential bonding

413–02 The standard UK system of shock protection is earthed equipotential bonding and automatic disconnection of supply (eebads). The principle is used to satisfy one or both of the basic international safety requirements to reduce the intensity or duration of an electric shock.

Bonding connections sometimes carry substantial network current – even with the local supply isolated. Always take care when disconnecting a bonding conductor for testing or service alterations.

The purpose of equipotential bonding is to maintain all simultaneously accessible parts within a zone at the same potential. It is not intended to carry fault current although it may well do so. Fault disconnection is achieved by earthing and that is a separate part of the exercise and normally relies upon circuit protective conductors.

Protective multiple earthing

Why earth at all?

A public electrical distribution system in the UK is earthed at the local transformer. If the supply was not earthed, it would eventually acquire a connection to earth at some stage by a fault on a consumer's installation. This would be harmless and might go undetected.

However, the system would now be earthed.

Subsequently a second fault to earth could arise on another installation. Current would immediately flow between the two independent earth connections and could cause fires at both places.

By making a positive earth connection at the transformer the supplier ensures that all earth faults can be detected and made safe by the operation of protective fuses, circuit-breakers or residual current devices.

Reliability of the earth–neutral path

Over the years, distribution networks and installations have become more complex, and with complexity there are increased chances of faults and open circuits. It therefore became technically expedient to earth the neutral conductor at different places around the network. This is protective multiple earthing.

The principle of pme is shown in Fig. 7.1. For clarity this is a simplified
Defs   single-phase system. The Wiring Regulations designate an installation using an earth derived from a pme network as TN-C-S. It is usually the consumer's

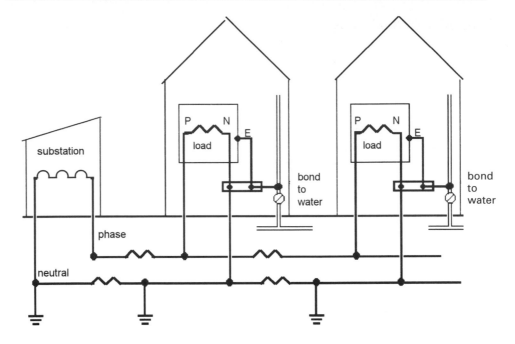

Figure 7.1 **The principle of pme.**

decision to use this earth. If the consumer prefers to use an earth electrode instead of the pme earth, the instalation is designated as TT.

Note the multiple earthing, and the fact that this may only be carried out on the supply network. The consumer is given three terminals: phase, neutral and earth. Within the installation the neutral and earth should *never* be interconnected. This is dangerous, contrary to the Supply Regulations and therefore illegal. If neutral and earth are accidentally or deliberately joined, an indeterminate current will flow which may be related to network conditions, not just the situation within the particular installation.

It will also be seen that all network conductors, including earth, have resistance. When current flows, resistance produces voltage drop. In the case of the soil, there will exist a potential gradient between earth electrodes. There could be a difference of a few volts between different 'earths' on the system. This is the reason why the introduction of 'earthy' extraneous conductive parts into an existing equipotential zone may create hazardous, or inconvenient, touch voltages.

Defs   Equipotential bonding of all services and extraneous conductive parts creates an electrical zone where everything within reach is effectively at the same potential. Isolated metalwork by definition is insulated from earth and is not an extraneous part. Isolated metalwork carries no potential and should not be bonded.

The equipotential zone may be extended to other related areas with the use of equipotential main bonding conductors.

Main bonding

547–02 Metalwork entering the zone from another area is sometimes already connected to the earthing on a different system. This could bring in a different earth potential. This metalwork must therefore be bonded locally to maintain the equipotential characteristics of the zone. Metalwork or other conductive media entirely within the zone is no problem.

Most of these items within the zone need not be bonded:

❐   Door handles,
❐   Window frames,
❐   Stainless steel sinks with plastic drainpipes,
❐   Suspended ceiling grids,
❐   Free standing steel tables or benches,
❐   Steel cupboards, shelving or racks.

Because of the use or proximity of electrical appliances or luminaires, it may be decided to *earth* some of the above excluded items. This is a subjective safety judgement and not considered to be an equipotential bonding issue.

Metallic water, gas and oil pipes entering a building must have a main bonding connection within 600 mm of the point of entry (see Fig. 7.2). In

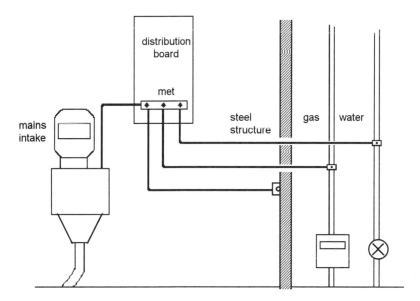

Figure 7.2   **Connections at main earthing terminal (met) must be accessible for testing.**

some cases it may be convenient to run one main bonding conductor for more
526–01 than one bonding connection. This is permissible provided that it is possible to
disconnect any single service without interfering with the bonding to other
services (Fig. 7.3).

Other items requiring a main equipotential bond are structural steelwork
and lightning conductors. A central heating system entering the building, or a
rising main would also need to be considered.

According to the Regulations, a bonding conductor need not always be a
543–02 cable. Some extraneous conductive parts could possibly be used. These must
have continuity that is permanent and reliable and have an appropriate
copper equivalent cross-sectional area. Suitable structural steelwork and
pipework may be used but not gas or oil pipes.

The use of extraneous metalwork for main bonding involves examination of
on-site conditions and an understanding electricity company inspector. Only
main bonding utilising cables is considered on the projects in this book.

Although the Regulations give minimum copper equivalent sizes of main

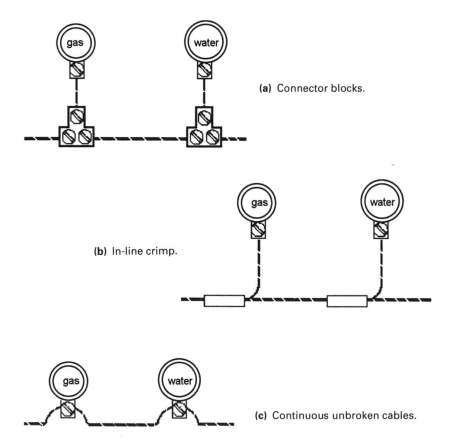

(a) Connector blocks.

(b) In-line crimp.

(c) Continuous unbroken cables.

Figure 7.3   **Three bonding methods using a single cable.**

Table 7.1 Minimum size of pme main bonding conductors.	
Copper equivalent cross-sectional area of supply neutral conductor (mm^2)	Minimum copper equivalent, cross sectional area of main bonding conductor (mm^2)
35 or less	10
35 up to 50	16
50 up to 95	25
95 up to 150	35
150	50

547–02 bonding conductors for pme systems as shown in Table 7.1, the local supply authority may have special requirements for main bonding.

Smaller sizes would be acceptable for TT and TN-S systems but as most installations will eventually be connected to a pme network it is suggested that the table should be used for all situations.

Single fault condition

413–02 An important factor to be considered is that the Regulations only consider a single fault condition. Bonding does not exist to clear faults on unrelated equipment which should have their own protective measures. The risk of a live conductor coming into contact with an extraneous part elsewhere should be eliminated by protective and functional insulation or with a circuit protective conductor.

It is only when a fault on equipment is compounded by a failure of insulation, or a circuit-breaker not operating, that an extraneous part can become dangerously live. Obviously bonding will help to remove a hazard, but this is not its primary function.

Supplementary bonding

547–03
413–02 Supplementary bonding is only normally required in areas of high risk, such as bathrooms. These applications are shown in projects in other chapters.

As with main bonding, the prime purpose of supplementary bonding is to ensure that all simultaneously accessible metalwork is at the same potential.

The Wiring Regulations give the opportunity to use supplementary bonding elsewhere where earth loop impedance is high. The condition rarely applies in small installations and requires special study, it is only mentioned in passing here. Nevertheless it is a good idea to supplement bonding wherever possible, for example by cross-connecting structural steel with a metallic switchgear. This will lower the total earth loop impedance.

It is not such a good idea to cross-bond to isolated metalwork such as windows or doors set in timber frames. Under certain conditions this bonding may transmit a potential to an otherwise safe component.

Circuit protective conductors

The subject changes now from bonding to earthing. This may be academic in that electric current will not take note of labels and differentiate between green–yellow cable routes, plumbing or structural metalwork. Fault current will be divided across all available paths in proportion to relative resistances in accordance with Ohm's Law.

Circuit protective conductors are the green/yellow wires connected to appliances or taken through conduits to outlets. It would be incorrect to call
543–01 them 'earth wires'. A cpc must be sized and installed to ensure that any earth fault current will operate the appropriate protective device safely and speedily.

Sizing of cpcs is related to:

❑   The size of live circuit conductors (i.e. phase and neutral),
❑   The rating of the protective device,
❑   Rapid disconnection within a designated time,
❑   The prospective fault current.

Note that the above does not include the normal load current. This will have been taken into account when selecting the phase and neutral cables and the associated protective devices.

This book does not enter into calculations for protective conductor sizes. The projects shown have been designed with suitable circuit protective conductors which will conveniently fulfil the requirements of the Regulations. In some circumstances the result is over-engineering and the designer may find commercial advantages in making different arrangements.

Steel conduit and trunking

Undoubtedly conduit has sufficient conductivity to be an effective circuit protective conductor. Unfortunately a problem arises regarding continuity at
543–02 joints, particularly where a coupler and bush is used at knock-out holes in sheet metal enclosures. It would be difficult to assert that every joint will give

permanent and reliable continuity over the course of the subsequent five years or more. Additionally, in the absence of definitive information, a full continuity calculation would need to be carried out for every joint in the installation.

In the circumstances, for the purpose of the projects in this book, it is recommended that a green–yellow insulated copper conductor be pulled into all conduits. The conductor should be the same size as the largest phase conductor in the conduit.

It is important to remember that this does not reduce the responsibility to ensure the continuity of the conduit system. This is necessary to earth the

526–01 conduit itself and constitutes a good reason to use insulated plastic conduit.

Steel trunking can be considered in the same light as conduit. Effective continuity is as good as the worst joint. If permanent and reliable continuity can be assured, there is no reason why the trunking should not be used as the sole circuit protective conductor. However, the designer must be assured that the cross-sectional area at joints corresponds to the formula given in the Regulations.

An easier and possibly more reliable solution is to install a separate copper circuit protective conductor.

Note that it is acceptable to use just one cpc and/or bonding conductor for a group of circuits. This conductor should be sized appropriate to the largest

543–01 circuit or bonding requirement. Tee-off connections should be made in a manner that allows modifications and disconnections to be carried out without breaking the continuity to other equipment. In-line crimped joints are recommended.

Steel wire armoured cable

Is the armouring on a cable adequate for 'earthing' purposes? This is a matter of terminology and, as will be seen, the armouring is never suitable as a main bonding conductor, but may be satisfactory as a cpc.

There are two methods of determining the size of a cpc:

543–01 ❑ Using the adiabatic formula in the Regulations, or
❑ Using the table of acceptable sizes.

Once again the issue of calculations is avoided in this book and only the second option is considered here. It must be emphasized that this method is probably unduly pessimistic and a more economical solution may result from detailed calculations.

Comparison of PVC and XLPE armoured cable

The physical sizes of insulation around similar conductor size PVC and XLPE cable are different. XLPE cable has a smaller diameter. The quantity of steel armour is therefore different and these cable types must be considered separately. The Regulations also give different insulation factors. Table 7.2 shows differences for a selection of cables. The table clearly illustrates the problems associated with use of armouring as a cpc without making calculations. The subject is complicated by the different maximum operating temperatures of the two types of insulation. These are 70°C for PVC and 90°C for XLPE.

As a general rule, on the commonly used smaller sizes of PVC swa cable, armouring is acceptable as a cpc. It may also be acceptable for some sizes of XLPE swa cable but calculations need to be carried out. These involve a knowledge of the prospective earth fault current on the project.

Continuity of cable glands

Regardless of any of the above considerations, it is essential to ensure permanent and reliable continuity at terminations of armoured cable. This is

Table 7.2 **Suitability of steel wire armour as a protective conductor.**

Cross sectional area of phase conductor (mm²)	No. of cores	Actual (mm²)	Cross-sectional area of armouring, minimum acceptable as cpc (mm²) OK	
PVC insulation				
16	2	47	36	✔
16	4 =	72	36	✔
25	2	61	36	✔
35	2	66	36	✔
35	4 =	85	36	✔
50	2	76	56	✔
95	2	123	107	✔
95	4 =	160	107	✔
XLPE insulation				
16	2	41	61	✘
16	4 =	49	61	✘
25	2	42	61	✘
35	2	62	61	✔
35	4 =	96	61	✔
50	2	68	95	✘
95	2	113	180	✘
95	4 =	140	180	✘

necessary to carry fault current if the sheath is penetrated and short circuited to the phase conductor.

526–01 The continuity of cable gland joints to steel switchgear enclosures is notoriously unreliable. Furthermore, enclosure sections fabricated from separate panels increase the chances of a high resistance fault path. To establish good continuity, an earth tag washer should be fitted to the gland and a cable linkage used across to the enclosure earthing terminal.

Information Technology equipment

Sec. 607 The Regulations devote a whole section to earthing arrangements for equipment which has a continuous earth leakage as part of the design.

Information Technology equipment includes most computers, data processing devices and other specialised apparatus such as life support systems in hospitals. For convenience, this chapter will refer to all such equipment as 'computers'.

545–01 A common feature of all computer equipment is the sensitivity to supply fluctuations and spikes on the network. These may damage transistorised hardware or corrupt data on software. To ensure a clean supply, computers are frequently installed on dedicated 'clean' circuits. They also have inbuilt suppression filtering circuitry. It is these capacitance bridge filters that leak unwanted components of the electricity supply to earth.

On small desktop computers the individual earth leakage is very small and of no great consequence. On large groups of personal computers or single items of mainframe equipment the total leakage may be substantial and introduces a potential hazard to personnel.

With good, reliable earthing on the consumer's installation, the earth leakage is taken away. However, if an open circuit occurs on the system earth, or a high resistance connection arises, everything connected to the faulty earthing system will acquire a potential from the filter circuitry. The fault may not occur suddenly; it may, for example, develop by way of corrosion at a joint.

Many computers have all-insulated enclosures; these would not cause a problem, but the fault potential will be transmitted to equipment with exposed metallic parts such as photocopiers or heating appliances.

Earth leakage currents

The installer will not usually have any information upon computer earth leakage and the supplier may be unhelpful. The Regulations suggest that

607–01   leakage below 3.5 mA may be ignored and a single desktop computer will be below this figure. The normal 13 A plug and socket connection is adequate for a single home computer.

Advice should be sought from the computer manufacturer for larger items of equipment or large office installations. Normal electrical test equipment may not give high frequency readings and cannot be used to measure computer leakage current.

❐ *Items of stationary equipment*

607–02   Where the leakage is estimated to be between 3.5 mA and 10 mA the equipment should either have a permanent connection to the supply or a BS 4343 'Commando' type plug should be used.

High integrity earthing

For large earth leakage conditions, the Regulations seek 'high integrity' earthing. This may be achieved with a positive earth connection with a minimum size of 10 mm^2. This is shown in Fig. 7.4(a).

The alternative is to supplement the normal cpc with an extra, dedicated, protective conductor.

❐   Where the item of stationary equipment has an earth leakage in excess of 10 mA the equipment should preferably have a permanent connection to the supply. Alternatively, a BS 4343 plug may be used, provided that the cable green–yellow protective conductor is supplemented by another protective conductor, with a minimum size of 4 mm^2 and connected by way of a separate pin on the plug.

Methods of supplementing the protective conductor are shown in Fig. 7.4(b) and (c). It will be seen that metallic conduit and cable armour or braid may be used, provided that connections are permanent and reliable. The same applies to metallic trunking.

Earth monitoring and isolated supplies

607–02 ❐   An earth monitoring system (see Fig. 7.5) incorporates a second protective conductor. This may be used as the supplementary earth provided that it is of the same standard as Fig. 7.4 methods.

❐   Where supply is fed from a double wound isolating transformer, a circuit protective conductor for the computer will be taken from the isolated secondary winding of the transformer. This connection should be supplemented to the same standard as Fig. 7.4 methods.

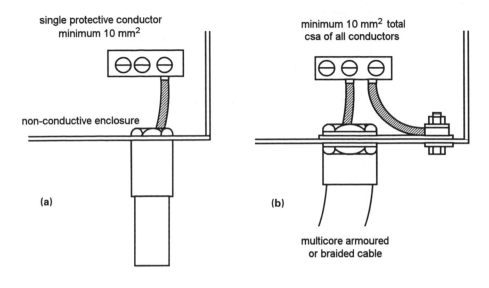

single protective conductor
minimum 10 mm²

non-conductive enclosure

(a)

minimum 10 mm² total
csa of all conductors

(b)

multicore armoured
or braided cable

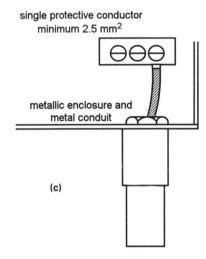

single protective conductor
minimum 2.5 mm²

metallic enclosure and
metal conduit

(c)

Figure 7.4 **High integrity earthing for computerized systems. (Phase and neutral conductors not shown.)**

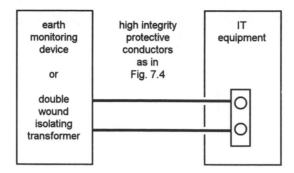

Figure 7.5 **Similar earthing requirements apply to safety systems. (Phase and neutral conductors are not shown.)**

Socket outlets for desktop computers

607–02 The Wiring Regulations refer to an installation with 13 A sockets which are intended to supply several items of equipment where the total leakage may exceed 10 mA. There are two ways in which the installer may provide high integrity earthing:

❑ One of the arrangements shown in Figs 7.4 or 7.5 should be applied, or
❑ A ring circuit may be used. This ring may only utilize single socket outlets and spurs are inadmissible.

Connections of protective conductors

Where more than one protective conductor on any of the systems described above is connected at a distribution board or equipment terminal block, separate terminal holes should be used.

Residual current devices

To avoid the loss of data from nuisance tripping, it is not advisable to apply residual current devices to computer circuits. Where these are specified for safety reasons, for example in schools, the total normal earth leakage of the equipment should not exceed 25% of the operating current rating of the rcd.

607–03   The Regulations require that TT installations incorporate a residual current device and the above requirement should be noted.

For TT installations it is also necessary to satisfy a formula which is intended to limit dangerous voltages arising on exposed conductive parts in normal service: *2 × earth leakage current (amps) × earth electrode resistance (ohms) must be less than 50.*

A typical example for a large office may give an earth leakage of 100 mA and earth resistance of 10 ohms.

$0.1 \times 20 = 2$

This obviously satisfies the requirement.

Chapter 8
Car Service Workshop

There are limitations with this standard design. Some hazardous locations require special expertise and are beyond the scope of this book. The complexities of these situations cannot be ignored and note should be taken of the conditions.

Standards and Recommendations

Health and Safety Executive document HS(G)41 deals with petrol filling stations. This requires the person certifying the electrical installation on these premises to be competent and experienced with BS 5345, *Installations in Potentially Explosive Atmospheres*. The standard deals with what is generally known as flameproof installations. Similar restrictions upon competence apply to areas in a car repair workshop which may be used for paint storage or spraying. Anyone not experienced with BS 5345 must either undertake training or call in an expert to produce a suitable specification.

This design for a car service workshop does not cover the forecourt area of petrol filling stations or BS 5345 situations.

An adaptable design

Provided that the above potentially hazardous areas are treated separately, the electrical installation in a vehicle repair workshop does not differ greatly from a typical industrial workshop. This design may be used as the basis for a small manufacturing unit or a warehouse with materials handling plant.

Motor vehicle repair premises

A suitable starting point for design is with reference to Health and Safety Guidance Note PM 37. This leaflet gives good advice upon electrical installations in motor vehicle repair premises. The document should be studied by the designer, installer and potential user of the installation.

Reference will be made to PM 37 and other H & SE guidance documents wherever appropriate.

Other interested parties

❏ *Fire authority*
Emergency lighting and fire alarms, particularly if the building is associated with a petrol service station.

❏ *Client's insurers*
Safety equipment and wiring systems.

❏ *Health and Safety*
Consideration of PM 37 and other advice.

❏ *Electricity company*
Check availability of a supply to suit the potential load and confirm the location of the intake position.

❏ *Dealer agency*
In the motor industry it is common for the vehicle manufacturer to specify systems and equipment to be used for approved dealer servicing.

❏ *Equipment supplier*
Full details of equipment are required to ensure that the electrical requirements will be suitable. Some of this equipment may be imported and not to British Standards.

Building structure and finishes

The layout of the building can be seen in Figs 8.1.and 8.2 which also give details of the electrical installation.

❏ *Floor areas (approx.)*
Workshop 320 m²
Offices, etc., 100 m²
The 20 m² spray area is excluded from the contract, except that provision is required for an electrical supply to the area.

Construction

❏ *Workshop*
Steel frame with brick walls to 3 m and lightweight wall cladding above. Pitched roof, 5 m clear to underside of roof frame girders, 7 m to apex. Floor concrete throughout.
The vehicle inspection pit is concrete, with steps down and drainage.

❏ *Office*
Fair-faced building block walls.
Ceiling height 2.5 m. Conventional plasterboard ceiling under timber beams.
The space above the office ceiling has a blockboard floor and is used for storage.

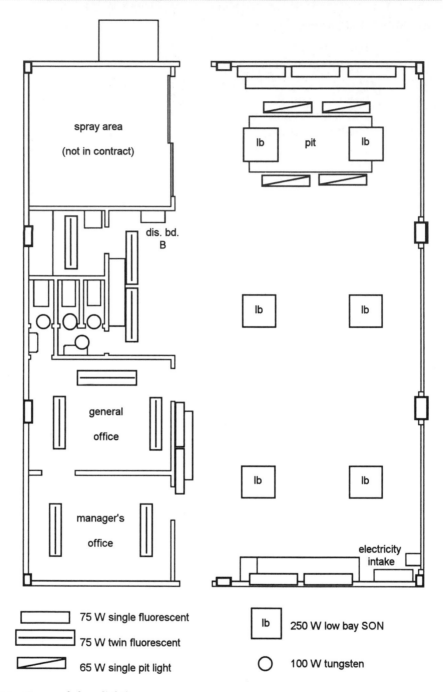

Figure 8.1   **Car workshop lighting.**

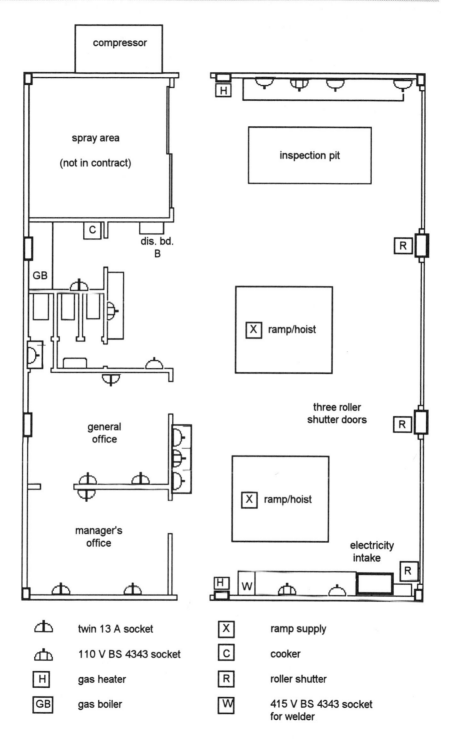

Figure 8.2 **Car workshop power.**

> ❏ *Space heating and hot water: gas*
> Workshop: two hot air blowers.
> Offices: central heating boiler.

Electrical requirements

A building layout plan is not sufficient to create an electrical design. Some specialist repair machinery requires dedicated supplies or special protection. A full schedule of equipment is required, listing:

- ❏ Lighting,
- ❏ Computerized equipment,
- ❏ Heating,
- ❏ Ventilation,
- ❏ Small tools,
- ❏ Large plant.

311–01 Early consultation with the client is necessary to establish the prospective loading.

Health and Safety Executive Guidance and Regulations

There are special hazards associated with this location where motor vehicles are repaired.

Health and Safety Guidance Note PM 37

- ❏ Steel conduit and trunking is recommended or PVC steel wire armoured (swa)cables.
- ❏ Where there is a risk of petrol spillage, no part of the electrical installation shall be less than 1 m above floor level.
- ❏ Switchgear must be accessible but located where it will not suffer damage.
- ❏ Lighting should be arranged to avoid stroboscopic effects from rotating parts.
- ❏ Fixed lighting beneath the vehicle inspection areas is preferable to portable handlamps.
- ❏ There must be no provision for portable electric tools or handlamps in inspection pits.
- ❏ Fixed luminaires in pits must be Zone 2 (sealed) units fitted flush with the wall at least 1 m above the floor of the pit.
- ❏ The use of handlamps should be discouraged. If they are used they should operate at 50 V.

471–15 ❐ Other portable electric tools should operate on a 110 V centre tap earth system. The use of reduced low voltage is no protection against fire and explosion from flammable vapours.

❐ Compressed air portable tools are preferable to electric tools.

❐ Industrial 240 V plugs and sockets to BS 4353 are more suitable than domestic BS 1363 types.

The installer should advise the client on the H & SE recommendation for 'Commando' type plugs. There is frequently user resistance to BS 4343 plugs and sockets for 240 V equipment.

(This scheme is based on BS 1363 13 A plugs with the BS 4343 alternative as an option.)

Wiring Regulations

❐ Every item of equipment must be of a design appropriate to the situation
512–06 in which it is to be used. The electrical installation must be of an industrial nature and, where appropriate, suitable for wet conditions.

❐ Any socket outlet which may reasonably be expected to supply portable equipment out of doors must have 30 mA rcd protection. Only the socket adjacent to the front roller shutter door is in this category but it would be a wise precaution to give rcd protection to all sockets.

Load assessment and maximum demand

After full consultations, a provisional list of electrical equipment has been drawn up in Table 8.1. A suitable format for a Project Specification is shown in Fig. 8.3.

Maximum demand load and diversity

This project calls for special consideration in terms of total maximum
313–01 demand. There is no relevant guidance. The designer must obtain manufacturer's data and rely upon experience.

Calculations and phase balancing figures for diversity will be approximate. Numbers have been rounded off.

Lighting

For current loading calculations, discharge lamp ratings must be multiplied by 1.8 to take into account control gear losses. An assumption of 100 W per outlet is made for tungsten lamps regardless of specification lamp size.

In the absence of other information, no diversity is being considered on discharge lighting and the tungsten load is discounted.

Table 8.1 Schedule of equipment for garage workshop.	
Offices	*Workshop*
Lighting Fluorescent 8 × twin 75 W 4 × 100 W tungsten	7 × single 75 W (benches) 4 × single 65 W (pit) 6 × 250 W SON (low-bay)
13 A 240 V sockets 8 × twin	6 × single 1 × single with rcd
16 A 110 V sockets (yellow)	4 × single
415 V welder (red), single phase and neutral	1 × 32 A single
Roller shutters, gas blowers and ramps, *3-phase and neutral*	8 × 6 A
Cooker and gas boiler supply, single phase	1 × 6 A 1 × 32 A
Provision for spray area, 3-phase and neutral	1 × 32 A
Compressor, 3-phase and neutral	1 × 16 A

Office lighting load

$$= 8 \times 2 \times 75\,\text{W} \times 1.8$$
$$= 2160\,\text{W}$$
This will be on one phase $= 9\,\text{A}$

Workshop lighting load

Benches $= 7 \times 75\,\text{W} \times 1.8$
$= 945\,\text{W}$

plus pit	$= 4 \times 65\,\text{W} \times 1.8$
	$= 468\,\text{W}$
plus low-bay	$= 6 \times 250\,\text{W} \times 1.8$
	$= 2700\,\text{W}$
Total lighting load	$= 4113\,\text{W}$

Assume that this will be spread approximately equally across three phases

$$\frac{4113}{240 \times 3} = 5.7\,\text{A/phase}$$

Welder

415 V 10 kVA rating. Full load must be anticipated across two phases
$$= 24\,\text{A}$$

Compressor

This runs on low load most of the time. 50% diversity can be allowed.

$$\frac{15\,\text{kW} \times 50\%}{240 \times 3} = 10.4\,\text{A/phase}$$

Gas blowers

2 × three-phase 0.75 kW fan motors. Full load must be anticipated.
$$= 2\,\text{A/phase}$$

Provision for spray area

In the absence of other information a maximum demand load has been assumed
$$= 30\,\text{A/phase}$$

Phase balance

It is essential to balance loads across three phases as far as possible. This exercise should consider the diversified current demands to obtain a balance on normal working conditions.

Table 8.2 gives a suitable arrangement. Note that at this stage, this does not necessarily indicate distribution board particulars.

Table 8.2 Proposed balance across phases using diversified load figures.

Phase		Amps
Red	Welder	24
	Office lights	9
	Works lights	5.7
	13 A sockets	20
	Gas blowers	2
	Spray area	20
	Compressor	10.4
	Ramp/hoists	—
	Roller shutters	—
		91.1 (Total)
Yellow	Welder	24
	Works lights	5.7
	13 A sockets	20
	Gas blowers	2
	Spray area	20
	Compressor	10.4
	110 V transformer	—
	Ramp/hoists	—
	Roller shutters	—
		82.1 (Total)
Blue	Works lights	5.7
	13 A sockets	20
	Gas blowers	2
	Spray area	20
	Compressor	10.4
	Gas boiler	—
	Ramp/hoists	—
	Roller shutters	—
	Cooker	25
		83.1 (Total)

Estimate of maximum demand

It will be seen that the estimate of maximum demand is very much a matter of experience. There is a tendency to overestimate high fixed loads which in practice only occur for short periods of time. Compressor and fan motors run 311–01 on low load most of the time and heaters have thermostatic regulation. The cooker is an unknown quantity, as are sundry kettles and room heaters.

On this project it is unlikely that the estimated maximum demand figure will ever occur for more than a few minutes. This information will be used to determine the size of the mains supply. If there is any doubt, it is worth showing the figures to the electricity company.

A 100 A three-phase supply will be suitable for this load.

What about a sub-main?

When the above data are studied it will be seen that the major loads are located some distance from the mains intake position. Long runs of steel wire armoured cables will be necessary for the spray area and the compressor. The cooker could also create cable sizing problems using PVC insulated and sheathed cable.

There are at least three technical and commercial disadvantages in running long final circuits:

❐ *Voltage drop*
525–01   This is usually the deciding factor for cable sizing.

❐ *Diversity*
This cannot be applied to final circuit cables which are selected with reference to the full load current and rating of the protective device. It may be acceptable to apply diversity to sub-main cables carrying several loads provided that the protective device is suitable.

❐ *Earth loop impedance*
The impedance along an extended small sub-circuit cable may exceed the limits for fault protection.

All three of the above factors apply to this scheme. The extreme example is the cable run to the compressor, a distance of some 50 m.

A convenient position for local sub-distribution is where it had been intended to locate a supply for the spray area requirements. A three-phase distribution board at this point will carry the circuits shown in Table 8.3.

Wiring systems

The workshop can be considered separately from the office situation.

Workshop

A layout such as this lends itself to a plastic conduit and trunking installation. These systems are easy to install but must be carefully located to avoid
522–06   physical damage. If this was a light industrial application, for example a

Table 8.3 Phase balance at sub-distribution board B.

Phase		Amps
Red	Spray area	20
	Compressor	10.4
	Office lights	9
		39.4 (Total)
Yellow	Spray area	20
	Compressor	10.4
	13 A Sockets	20
		50.4 (Total)
Blue	Spray area	20
	Compressor	10.4
	Cooker	25
		55.4 (Total)

Diversified loading figures have been taken from Table 8.2. The phase balance at the mains distribution would be unchanged.

clothes manufacturing workshop, non-metallic wiring systems would have advantages and would have been recommended. With this current project, electrical safety relies upon protection of cables in extreme circumstances. Guidance Note PM 37 recommends the use of steel conduit and trunking or swa cables for places where motor vehicles are serviced. It would be irresponsible to ignore this guidance.

Steel conduit and trunking

Trunking is suggested for major groups of cables for lighting and sockets around the building. A suitable clear route will usually be found on the walls at about 3–4 m above floor level; 415 V, 240 V and 110 V single-core cables may be mixed, provided that standard 300/500 V minimum rated cables are used.

528–01

In the workshop location, ambient conditions will always be reasonable; therefore, if cable runs are surface mounted and never buried in any form of thermal insulation, there will be no temperature or insulation derating factors to be applied. The sizes given in Table 8.4 have a good margin of tolerance in sizing to the point that grouping factors may also be discounted.

523–04

Table 8.4 Cable sizes related to mcb ratings.

Circuit	Full load (A)	mcb (A)	(type)	Cable type reference	Size (mm^2)	Max. length (m)
Distribution Board A						
Sub-main to Distribution Board B	Assume 60	63	3	XLPE–swa (4-core)	10 or 16	25 40
Ramps	2.8	6	3	PVC-swa (4-core)	1.5	50
Works lights (3 circuits)	6.0	10	3	singles 6491X	1.5	30
13 A sockets	32 A ring or 20 A radial	32 20	2 2	singles 6491X singles 6491X	2.5 2.5	66 30
Roller shutters	1.0	6	3	PVC-swa (4-core) or singles 6491X	1.5 1.5	50 50
Welder	24	32	3	singles 6491X	6.0	40
Gas heaters	2.0	6	3	PVC-swa (4-core)	1.5	50
Distribution Board B						
Office lights	9.0	10	2	T & F 6242Y	1.5	55
Office sockets	32 A ring or 20 A radial	32 20	2 2	T & E 6242Y T & E 6242Y	2.5 2.5	66 66
Cooker	30	32	2	T & E 6242Y	6.0	40
Gas boiler	2.0	6	2	T & E 6242Y	1.5	50
Compressor	10.4	16	3	PVC-swa	4.0	50

Steel conduit will be taken from the perimeter trunking to lighting fittings and socket outlets.

Steel wire armoured cable

Undoubtedly swa cable is an economical method of taking single services to isolated equipment. The schedule therefore gives this as an option for door roller shutter mechanisms and the vehicle ramps.

It must be remembered that swa is not a flexible cable for unsupported connections to machinery. For vertical drops, support at 500 mm maximum is essential with care taken to ensure that no stress is put on to glanded entries 522–08 into enclosures. Two methods are suggested for vertical supply drops from roof girders down to fixed equipment in workshops:

❐   Clip to Dexion or similar vertical supports.
❐   Terminate the swa at high level and make the drop in flexible cable.

Horizontal runs of swa along walls should be clipped at 400 mm maximum intervals, or laid on a continuous support, such as a structural steel girder or cable tray.

Office

The office construction appears to be of an 'industrial' nature with plain building block walls. There is no cavity or plaster cover to hide PVC-sheathed cables, but the ceiling is conventional plasterboard fixed to joists with a floor above. This is a difficult location in which to install conduits.

The choice is either a complete surface-mounted conduit system throughout the office and toilet area, or sheathed cables concealed in the ceiling and enclosed in conduit for wall drops. In either case, plastic conduit is appropriate, the alternative being mini trunking for sheathed cable drops to lighting switches and sockets.

Arrangement of circuits

Using the schedules produced for balancing loads at two positions, distribution board circuitry can be planned. To size each sub-main and final circuit mcb, full load current must be used. This in turn will determine cable sizes.

Distribution boards

Three phase type B mcb distribution boards are suitable for both locations. Board A will carry the whole load and feed board B (see Fig. 8.4).

Distribution Board A: 125 A main switch, 24 outgoing ways
Distribution Board B: 125 A switch (optional), 12 outgoing ways.

Project Specification
BS 7671

Garage workshop

Name...................... Location.......................

Reference................ Date.....................

Three phase 240 V 50 Hz. TN-C-S.
Supply fuse 100 A BS 1361 or BS 88

PFC less than 16 kA. Earth loop impedance less than 0.35

100 A TP main switch

Circuits	Rating (A)	Cable size (mm^2)	Max length (m)	Lights/points g = gang
Main distribution board A 125 A isolator 24 ways				
Lights wkshp 1	10	1.5	30	6
Lights wkshp 2	10	1.5	30	6
Lights wkshp 3	10	1.5	30	5
Ramps 3-phase	6	1.5	50	3
13 A sockets wkshp	32	2.5	66	3 x 1g
13 A sockets wkshp	32	2.5	66	4 x 1g
Roller shutters 3-phase	6	1.5	50	3
Welder socket 2-phase	32	6.0	40	1
110 V transformer	10	2.5	30	4 x 1
Gas heaters 3-phase	6	1.5	50	2
Supply to dis bd B	63	16	40	-
Sub distribution board B 125 A isolator 12 ways				
Lights office	10	1.5	55	12
13 A sockets office	32	2.5	66	8 x 2g
Cooker	32	6.0	40	1
Compressor 3 phase	16	4.0	50	1
Spray area provision	32			
Gas boiler	6	1.5	50	1

Figure 8.3 **Project specification for garage workshop.**

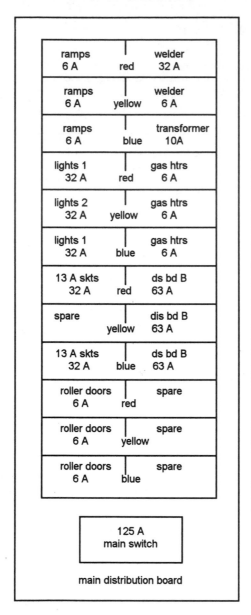

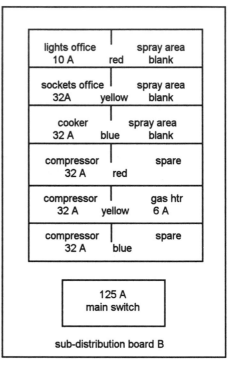

Figure 8.4 **Distribution board circuits.**

Cable sizes

Sizing must take account of mcb rating, any applicable derating factors, earth loop impedance and voltage drop on full load. With the workshop layout shown, voltage drop is the most onerous. Suitable cable sizes are shown in Table 8.3. Maximum lengths of run should be noted if the design is used for a different layout.

Isolation and switching

Sec. 460

Main switch, Distribution Board A

This must be three pole. Switching of the neutral is not required.

Distribution Board B

It is not essential to switch this board unless there is a particular requirement.
461–01 Isolation can be carried out by locking off the 63 A mcb at Board A.

Machinery

- Compressor,
- Gas heaters,
- Ramps,
- Roller shutters,
- Welder.

All these items require:

- Switching for isolation,
- Switching for mechanical maintenance,
- Emergency switching.

Isolation and switching for mechanical maintenance could be achieved by locking off the appropriate mcb at the distribution board. The requirement for emergency switching must be immediately accessible to the user of the equipment.

All three functions are best combined in an appropriate on-load isolating switch adjacent to each item.

Cooker

This requires all three functions listed above. The most convenient method is with a conventional cooker control switch of the type without an integral socket outlet.

Gas boiler

This requires an isolating switch adjacent to the boiler. The switch will supply the boiler system including controls and valves.

110 V transformer

This only requires switching for isolation which could simply be a lock-off arrangement on the mcb at the distribution board. A switch adjacent to the transformer may be more convenient and would provide a functional facility to switch off the 110 V system at night or at other times when it is not required.

Earthing and bonding

Standard earthing requirements apply as shown in Fig. 8.5, but there are some possible variations.

Main earthing terminal

The 16 mm^2 main earthing conductor from the supply pme earth is taken to 547–02 the main earthing terminal at Distribution Board A.

Main bonding conductors at 10 mm^2 are required from the main earthing terminal to:

❑   Main water stop cock,
❑   Main gas stop cock,
❑   Structural steelwork.

Protective conductors at Distribution Board B

Provided that all main bonding is carried out at Distribution Board A there are no bonding requirements at Distribution Board B. Circuit protection is the requirement.

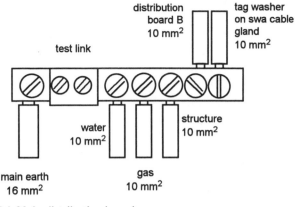

(a) Main distribution board.

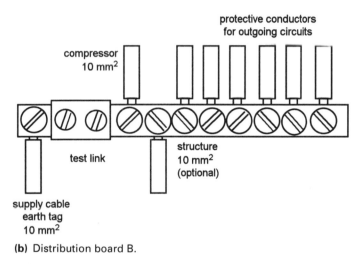

(b) Distribution board B.

Figure 8.5 **Earthing bar connections.**

❏ The earthing terminal within this remote distribution board provides a marshalling terminal for local circuit protective conductors.

❏ The swa armouring may not be adequate as a circuit protective conductor (see Chapter 7). There are often problems with XLPE cable. It may be preferable to use a cable with a separate green–yellow protective conductor.

❏ To improve the earth-loop impedance it is a good idea to make an additional bonding connection to the structural steelwork at this sub-distribution board. This could be achieved simply by bolting the metallic enclosure to a steel stanchion.

Armoured cable glands

It is important always to fit an earth tag washer to the cable gland whether or not the cable armouring is used as a protective conductor. Even if a separate cpc is run, it is necessary to have permanent and reliable continuity to the 526–01 armouring. The tag washer should be linked to the earthing terminal in the distribution board or other enclosure. It is not good enough to rely on continuity through the metal box.

Steel conduit and trunking

To ensure permanent and reliable continuity it is advisable to install a separate insulated cpc through conduit and trunking but this does not reduce the obligation to make good joints in the steel system.

Chapter 9
Circuits

Terminology

It is essential to use the correct terminology when describing electrical installation subjects. This is of particular importance in this chapter for the description of conductors.

The conductors on a single-phase supply are correctly known as:

❏ *Phase*. This is coloured red on hard wiring or brown on a plug terminal.
❏ *Neutral*. This is coloured black on hard wiring and blue on a plug terminal.

Both of the above are known as live conductors.

❏ *Circuit protective conductor (cpc)*. This is always green–yellow.

A cpc should not be confused with a bonding conductor which is also coloured green–yellow.

Colours of three phases

Three-phase supplies have three-phase conductors which the Regulations require to be coloured red on outgoing ways of single-phase final distribution

514–06 boards. The colours blue and yellow may only be used for three-phase circuits and, where desired, on the supply side of single-phase distribution boards.

Strictly speaking it is wrong to use yellow and blue conductors on two-way switching circuits. Regardless of this, in many areas, wiring with three-coloured multicore cable has become a local convention which simplifies identification of wires at the switch. Care must be taken to ensure that this practice is never carried out on an installation with three-coloured cables for three-phase circuitry. A blue or yellow two-way switch connection could be taken as phase colours. It is also contrary to Regulations to wire single-phase lighting and socket-outlet circuits in 'phase' colours. This is often a feature of large installations and there is logic in the identification of circuit phases.

Where the above irregular colour codings are used, the completion

App. 6 documentation should be marked to show appropriate deviations from the Regulations.

Conventional circuits

IEE Guidance Notes for the Regulations now describe 'conventional' circuits. These are based on previous ideas for standard circuits and generally apply to power sockets.

Lighting circuits

The industry has developed its own conventional circuitry for applications such as lighting and motor control. Chapter 2 showed that there are alternative arrangements for wiring domestic lighting. For example, systems using twin and earth cable may be looped at the light position or at the switch. The choice will be at the convenience of the electrician. Such elementary matters are not usually contained in specification documents.

Where conduit wiring and single-core cables are used, another convention is to loop phase conductors from the switch and neutral conductors at the light.

It has also been known for a room light to be supplied from power circuits through fused connector units. This practice may be recommended for hotel rooms. One 30 A circuit can supply all requirements in one or two rooms.

Induction

Incorrect use of single-core lighting cables may introduce inductive nuisance for hearing-aid users or sensitive data processing activity. Chapter 2 showed a method of arranging two-way switch wiring to ensure that inductive effects are 331–01 cancelled out. With the increase in use of electronic equipment and building management systems, this subject is becoming increasingly important.

Socket outlet circuits

The circuit conventions for power outlets are currently undergoing scrutiny. It should be remembered that the standard 13 A arrangements related to floor area were intended for domestic installations. They have little relevance to commercial, industrial or other larger buildings.

Changing methods

Electrical designers and electricians have become accustomed to standardised wiring methods which work on rule-of-thumb principles. This is good to the extent that everyone in the business understands circuitry conventions. One installation is very much like any other.

Whilst it makes sense to use tried and tested methods, we should not be so blinkered as to think that nothing will ever change. In recent years there have been tremendous changes in the utilization of electricity, and equipment manufacturers have introduced some very sophisticated products. Switchgear and control equipment designers have taken up advanced technology associated with new materials and computer processing.

At a very basic level, electrical installers may be in danger of letting the world pass them by. Electrical appliances and equipment have changed on both domestic and commercial premises. Loadings are not what they were a few years ago, yet circuitry remains unchanged.

Ring main obsolescence

Defs
The one wiring system common to all types of installation in the UK is the 30 A ring main, or, using correct terminology, the ring circuit.

Strangely enough, the ring circuit will not be found in Europe or America or Japan, or anywhere outside the old British colonial sphere of influence. Have we got it right, or is this another of the UK's outdated insular customs?

History of ring circuits

In the late 1940s, the government of the day set about raising standards for postwar housing. Electricity was becoming a basic necessity and old methods of wiring concentrated on lighting and one or two power appliances in the kitchen. Few homes had anything more than a radio and a domestic iron, both of which could be plugged into a lighting two-way adaptor.

Central heating was unknown in the average home and was not taken into account in the design of a revolutionary domestic electrical power distribution system: the ring main.

The principle of the ring main system was that a householder would have one, or even two, electric heaters which could be taken from room to room and plugged in to a convenient socket. This facility for limited space heating plus the loading of an electric kettle in the kitchen was seen to require a 30 A supply.

Rather than run a 30 A cable to every outlet an idea was developed to use a 20 A cable in the form of a ring with 13 A fused plugs. This permitted the connection of a 3 kW load at any socket outlet. It was assumed that the average small householder with less than 1000 ft^2 (or 100 m^2) of floor area would not have a sufficient number of appliances to overload the ring.

Times have changed

This was all very good for homes in the 1940s and 1950s but bears no

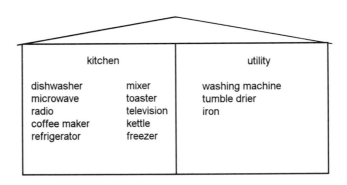

bedroom 1	bedroom 2	bedroom 3
2 bedside lamps clock radio electric blanket teamaker electric iron hair drier	2 bedside lamps transistor radio hair curler electric blanket games computer	bedside light transistor radio electric blanket model railway
lounge	dining room	hallway and study
television video music centre 3 table lamps	plate warmer 3 table lamps radio	vacuum cleaner answer phone computer table lamp

Figure 9.1 **Small diversified loads outside of the kitchen.**

kitchen		utility
dishwasher microwave radio coffee maker refrigerator	mixer toaster television kettle freezer	washing machine tumble drier iron

Figure 9.2 **Heavy loading requires a 30 A circuit.**

relation to domestic requirements at the end of the 20th century, as shown in Figs 9.1.and 9.2. Furthermore, the allocation of 30 A rings to floor areas was never intended to apply to commercial situations, even as they existed in 1950. In today's high-tech office the use of the 100 m^2 formula is a nonsense. It all depends on which particular 100 m^2 area is being serviced.

There are good reasons for considering alternative wiring systems:

□ Domestic and commercial consumers have a multitude of low-current appliances. Many more sockets are required in a new installation and flexibility is needed for future alterations and additions. Extending or breaking into a ring circuit is not a straightforward exercise.

□ A dedicated 30 A circuit is recommended by the Regulations for the domestic kitchen/utility room, or two 20 A circuits where these are separate areas.

□ Unless thorough testing is carried out on a new or modified ring circuit, wiring faults may go undetected and invalidate the basic safety principles of the system (see Fig. 9.3).

□ The ring circuit does not lend itself to separate control of groups of outlets for consumer convenience.

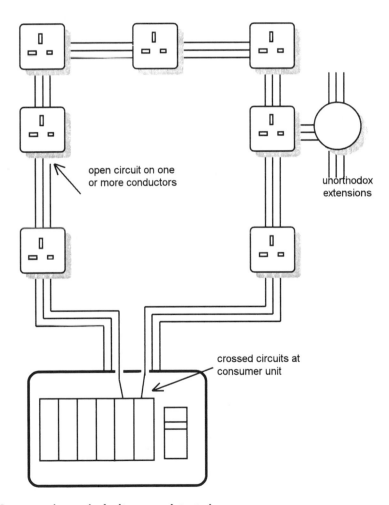

Figure 9.3   **Common ring main faults are undetected.**

◻ There is international movement in the direction of a universal 16 A unfused plug. Even if this takes another ten years to become established it will still be introduced within the life of today's electrical wiring.

It is possible to change a ring into two 16 A or 20 A radial circuits but there are better alternatives:

◻ It is now permissible to use foreign standard plugs and wiring systems for UK installations. These may have superior flexibility and economy although perhaps at the expense of quality. A better system for British plugs and sockets is essential.

Alternative methods

Multiple plug sizes

The old round pin plug systems used three different plug sizes, each of which had to be carried on separate circuitry. In theory, 5 A and 2 A plugs would have the capacity to handle lighting and low current appliances. 13 A or 15 A plugs would be needed for portable heaters and kitchen appliances.

This would contravene the most essential feature of any system, that of being able to use one type of plug and socket throughout an installation.

Radial circuits

IEE Guidance Notes show radial circuits in the conventional circuit arrangements. There is no detail showing exactly what a radial circuit looks like, but most people seem to imagine that it is a single run of cable connected, in turn, to successive sockets in a long line. This arrangement has limited application, especially if the guidance restriction to 50 m^2 floor area is literally applied.

Introducing the tree

Designs for projects shown in other chapters are based on conventional ring circuitry, although the floor area limitations are not always applied. These arbitrary figures are not in the Wiring Regulations and, as shown in the Guidance Notes, only apply to the domestic scene.

The tree systems described here comply with the Wiring Regulations requirements with regard to current-carrying capacity in typical situations where no cable derating factors apply. They may be applied as appropriate to all types of installation.

As with any other circuitry, the designer or installer must make a judgement on loading and diversity for each situation. It could be said that the tree is a variant on a conventional radial circuit but the name is more descriptive of the arrangement.

20 A tree

A 2.5 mm^2 cable is adequately protected by a 20 A mcb and there is no reason why the system should not branch off in any direction from socket or joint box terminals (see Fig. 9.4).

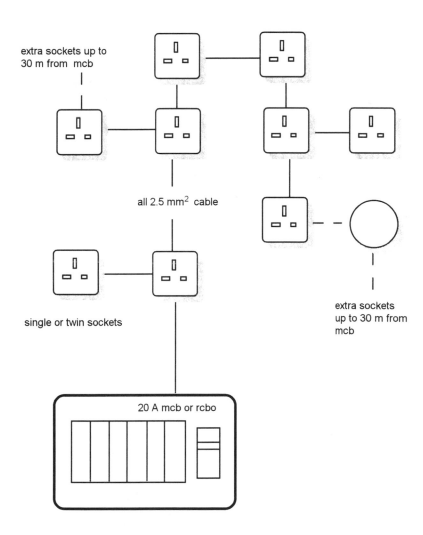

Figure 9.4   **A versatile 20 A tree system.**

Domestic

Outside of the kitchen, domestic electrical loading is made up of many low-current appliances with perhaps one portable heater. When diversity is applied, a continuous load in excess of 5 kW is unlikely.

In the absence of any official guidance the installer must decide upon areas to be covered by one tree. One tree circuit may be designated to two or more suitable rooms. Table 9.1 gives some ideas.

Commercial and similar

The 20 A tree may be applied taking into account the following limiting factors:

❑   Diversified current loading must not exceed 5 kW in the area designated to one circuit.
❑   Earth loop impedance to the most remote socket on any branch must not exceed regulatory Z_s limits. For practical purposes the length of run is of no consequence if the circuit is protected by a 30 mA rcd.
❑   Volt drop to the most remote socket on any branch should not exceed 4% on full load. On a typical pme supply, this requirement will be found to be

Table 9.1 Suggestions for loading limits on domestic tree circuits.

Tree system	Typical application for one circuit
20 A mcb 2.5 mm² cable Maximum run from mcb to furthest socket = 30 m	Living room and dining room or three bedrooms or sheltered flatlet All the above exclude kitchens but may include stairs and landings.
	A kitchen may have its own circuit provided there is a separate laundry or utility room.
30 A/32 A mcb 4.00 mm² cable Maximum run from mcb to furthest socket = 45 m	Kitchen and utility. 2.5 mm spurs as with a conventional 2.5 mm² 30 A ring.

more onerous than the earth loop impedance. A distance of 30 m to the extremity of a branch will handle the fully loaded condition.

32 A tree

A 4.0 mm^2 cable will be adequately protected by a 32 A mcb. Cables may branch off in any direction from socket or accessible joint box terminals.

Domestic

One 32 A tree circuit may be applied in a kitchen which encompasses domestic laundering equipment.

Commercial and similar

The 32 A tree may be applied taking into account the following limiting factors:

- ❑ Diversified current loading must not exceed 8 kW in the area designated to one circuit.
- ❑ Earth loop impedance to the most remote socket on any branch must not exceed regulatory Z_s limits. For practical purposes this length of run is of no consequence if the circuit is protected by a 30 mA rcd.
- ❑ Volt drop to the most remote socket on any branch should not exceed 4% on full load. On a typical pme supply, this requirement will be found to be more onerous than the earth loop impedance. A distance of 45 m to the extremity of a branch will handle the fully loaded condition.

Switching and control

One very interesting aspect of a tree system is the ability to switch groups of sockets. It is often useful to have a switch by a door to operate table lamps. In other conditions time switches or security detectors could control a bank of sockets. None of this is easy to arrange with a ring circuit.

As a final illustration of possibilities, Fig. 9.5 gives some ideas for two rooms in a domestic situation.

Comparison of systems

There is no universal solution to every wiring situation but in many cases a 20 A tree system may be a suitable alternative to a 32 A ring. The 30 m

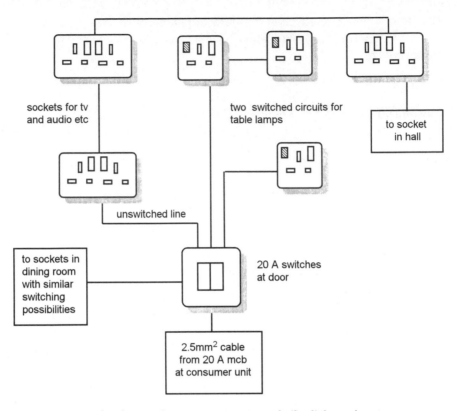

Figure 9.5 **Tree system for downstairs rooms. (Note use of pilot light sockets.)**

maximum length of run to the furthest socket on a branch compares favourably with a 70 m limit on the total length of a ring in similar circumstances.

32 A ring

Disadvantages

- ❑ Untested installations or alterations may function with undetected faults,
- ❑ Unsuited to modern needs,
- ❑ Inflexible for alterations and additions,
- ❑ In many cases overrated for intended loading,
- ❑ Will not easily convert to a future 16 A system.

Advantage

- ❑ 32 A capacity for large loads.

20 A tree

Disadvantage

❑ Limited to 20 A.

Advantages

❑ Open circuits are easily discovered,
❑ Crossed circuitry is impossible,
❑ Additional sockets are easily added,
❑ Simple conversion to a future 16 A system,
❑ Possibly up to 20% saving in copper,
❑ Switching or other controls can be built in to groups of sockets.

Composite circuits

There are some applications more suited to a 32 A ring circuit for large loads. A 32 A tree with 4 mm^2 cables may save on labour but not on material costs. Surprisingly two 2.5 mm^2 cables are cheaper than one 4.0 mm^2 cable.

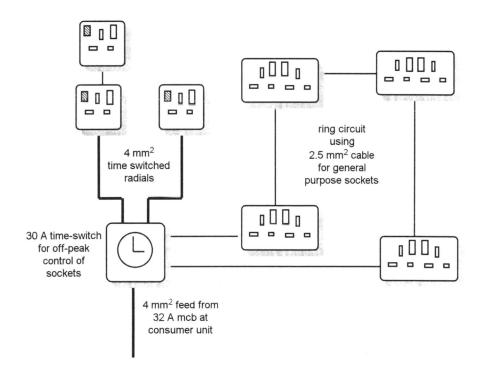

Figure 9.6 **30 A radial-ring circuitry for kitchen.**

It is acceptable to use a combination 32 A configuration combining the advantages of each system. One arrangement is shown in Fig. 9.6 but ideas like this may have greater application for commercial installations.

Chapter 10
Farming and Horticulture

For convenience the word 'farm' will be used generally to describe locations covered by this chapter. This description will include all types of agricultural and horticultural conditions including:

- ❑ Animal husbandry units and zoos,
- ❑ Stables, intensive breeding units, catteries and kennels,
- ❑ Commercial greenhouses, market gardens and garden centres.

But this chapter does not cover farm houses, offices or similar places for human accommodation. These should comply with standard Regulations.

The chapter is different from other chapters for two reasons:

- ❑ Farming projects often involve installations that are not connected with a pme supply network. Even if the electricity company has pme facilities in the locality, it may be undesirable to utilize them on a farm.
- ❑ It is difficult to devise a sample scheme that will take into account all the possible operations and conditions. There is no such thing as a 'typical' farm.

It would appear that the best method to deal with electrical installations in agriculture and horticulture is to consider each part of the installation in turn. Ideas will be given for design techniques, materials, safety and economy.

Sec. 605 The Wiring Regulations show the importance attached to farm electrics by devoting a separate section to special agricultural conditions.

Special Regulations for agricultural and horticultural locations supplement 600-01 or modify the general Regulations. They are not alternatives.

Why farms are different

Environment

By their nature, agricultural and horticultural locations suffer from extremes of climate coupled with the existence of mud, dust and agro-chemicals. These, and the presence of animals, may affect the choice of switchgear and wiring systems. Risks include fire and explosion. Vermin may cause problems.

Livestock

Farm animals are especially sensitive to electric shock. Fifty volts can kill a cow or horse standing on a wet surface. Even if the animal survives a shock, the distress caused may create a panic with disastrous results to both man and beast. An animal will remember the slightest shock 'tingle' and will refuse to re-enter a pen or stable where it encountered the problem.

In many intensive rearing situations the continuation of electrical heating or ventilation is critical to the survival of livestock. Reliability of equipment is very important.

Equipotential zones

Chapter 7, Earthing and Bonding, explained soil voltage gradients and voltage drop in cables.When these are considered it will be seen that it is impossible to create a huge equipotential area across a farmyard or between separated horticultural hothouses.

Equipotential bonding to each and every piece of extraneous metal is impracticable and in any case, bonding conductors have their own resistance, and with resistance comes volt-drop. Furthermore, as has been stated, farm animals are super-sensitive to electric shock.

For general non-hazardous locations, the Wiring Regulations accept a maximum of 50 V *touch voltage* as a safety norm. This is unacceptable on a 605–05 farm, with the added human and animal risks, where all calculations involving touch voltage use a figure of 25 V. It will be seen that farm electrical instal-
605–06 lations are special and should not be undertaken by inexperienced installers. It is not a do-it-yourself location.

This chapter can only offer guidance and short cuts to installers who fully understand the hazards – and have the appropriate insurance cover.

Earthing systems

Because farm installations may have earthing problems, we shall briefly review conditions that are explained in more detail in Chapter 7.

An installation using the pme earthing facility is described by the initials TN-C-S. Within the consumer's installation, the neutral is classed as a live conductor and must be separate from earth. It will have a common termination at the mains position.

Where either the supply earth is unsuitable or the provision of safe equipotential zones is unreliable, the supply authority may not give an earthing facility to the consumer. This is a TT system. The consumer must

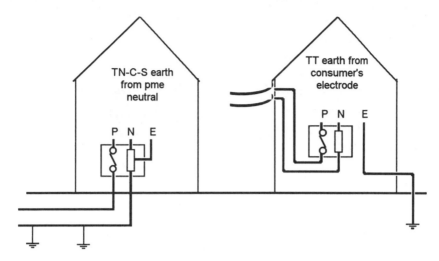

Figure 10.1 Usually TN-C-S is an underground service and TT is overhead, but not always.

provide an earth electrode to establish a local, safe equipotential reference (see Fig. 10.1).

In some instances the consumer who is given a pme supply may be advised not to use the supply earth. In other situations it will be found to be impossible to use this earth and comply with earthing loop impedance requirements for some or all of an installation. Under such conditions the consumer will use an earth electrode for the local earth facility. This again is a TT installation.

It can be seen that some parts of a consumer's installation may be TN-C-S and others TT.

It is not unusual for the pme electricity supply to be taken to a farmhouse which has a TN-C-S installation; this system in turn supplies sub-mains feeding outbuildings and barns which have local earth electrodes and are therefore TT installations. The earthing between these systems should not be interconnected.

Special earthing regulations on farms with TT systems

These apply to any installation where there is no connection to the supply company's earth, or no earthing facility is given.

☐ Every protective conductor (green/yellow cable) must be connected to a
542–01   common earth electrode via the main earthing terminal. If there are
    several protective devices in series, each section of the installation may be
413–02   connected to a separate earth electrode (see Fig. 10.2).

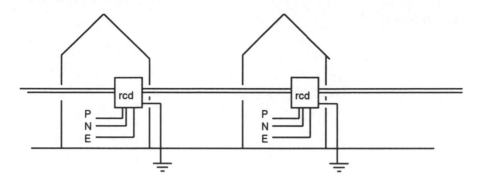

Figure 10.2 **An earth electrode for each building.**

413–02 ❐ A residual current device is the preferred protective device for shock protection. It is the only device that will be considered for shock protection in this chapter. Every TT installation will have at least two rcds. Fuses and circuit-breakers will be used for overcurrent protection.

605–05 ❐ A formula is given in the Regulations for the maximum resistance of an earth electrode. This relates to the 25 V maximum touch voltage. Table 10.1 gives typical maximum values.

Table 10.1 **Minimum earth electrode resistance for farms.**	
RCD operating current	*Max. resistance of total earth path*
I∆n (mA)	*cpcs + electrode* (ohms)
30	833
100	250
200	125
300	83
500	50

Warning

An earth electrode resistance in excess of 100 ohms must be considered to be unreliable and usually unacceptable.

Even readings above 50 ohms are suspect.

Earth electrodes

542–02

It is conventional practice to drive in an earthing spike. This is acceptable provided that a British Standard type electrode is used with the cable connection properly protected in a suitable enclosure. A length of galvanized

511–01 water piping is not satisfactory and an exposed termination will be easily damaged.

The electrode and its connections must be located in a position where it will not dry out, freeze or be subject to corrosion. This is especially important where chemical or slurry contamination may arise. It must be remembered that the shock protection features of the installation depend upon the effectiveness of this earth electrode current path. The installer carries full responsibility for future safety.

Alternative electrodes

542–02

Structural steelwork

The steel frame of a barn may create an excellent earthing electrode. The relevant requirements are that a suitable resistance be achieved in a permanent and reliable manner.

Concrete reinforcement

Steel reinforcement in *pre-stressed* concrete columns or beams does not constitute a reliable earth electrode regardless of any initial test reading.

Reinforcement metalwork in poured concrete, such as floor areas, may give good earthing provided that connections are reliable and can be made accessible for future testing.

Earthing grids or plates

Where the nature of the soil gives a high or unreliable resistance, consideration should be given to the use of buried earthing mats. This is a specialist subject and depends upon local soil conditions. Consultation with the supply company is recommended.

Pipework of other services

Although there may be a requirement to bond to water, gas or oil supplies, these must not be used to provide earthing facilities. The increasing use of plastic service pipes along some parts of a system will introduce shock hazards if isolated metallic sections are used for earthing.

Bonding

Main bonding

542–04 Standard arrangements apply whether the system is TT or TN. A main earthing terminal (met) must be established at the mains position. This connects the system earthing conductor to all main bonding conductors.

547–02 All incoming services must have main bonding and special note taken of unusual conductive pipework such as that for compressed air, fuel oil or water from a well. Where plastic pipework is used for water, a main bonding connection should be made on the metallic stop cock. After this, no further main bonding should be necessary within the equipotential zone covered by the earthing facility. If the water service is taken into another equipotential zone, main bonding will need to be repeated at each point of entry.

Milk pipes usually consist of sections of stainless steel with plastic connections. These should not be connected to any bonding or earthing system.

Supplementary bonding

547–03 This is required in places where humans or livestock may touch exposed or extraneous metalwork. For practical purposes, earthy metalwork normally

605–08 constitutes no major electrical hazard but, on a farm, slightly different potentials sometimes arise as a result of soil and conductor resistances. Animals will respond to this condition which is most apparent on pme networks.

Supplementary bonding requires that external conductive parts of the electrical system and all extraneous (earthy) conductive parts should be cross-bonded. This includes floors which may become wet in buildings where animals are housed. Typical locations are milking parlours and pig-rearing pens.

Early consultation is essential on all new farming developments to ensure that metallic grids are provided where necessary in concrete floors.

Common sense in bonding

A certain level of common sense and rationality is required when carrying out this supplementary bonding exercise. Metalwork such as wall mounted mangers and troughs will be at the potential of the building. Supplementary bonding is superfluous. Similarly, moveable barriers and isolated shelves or racking do not constitute a hazard.

It may well be that unnecessary supplementary bonding in these circumstances actually *introduces* a hazard.

A sensible approach to supplementary bonding is to cross-bond only where a positive hazard can be identified and to avoid bonding in all other circumstances.

Residual current devices

Discrimination

531–02 Before considering the various requirements for an rcd on farms, the factors related to discrimination should be re-examined. A problem may occur when there is more than one rcd in series with the supply to equipment or an outlet.

Discrimination requires that the protective device nearest in line to a fault should operate before up-stream devices. Thus, if there is a 30 mA protected socket in a workshop, this should trip before a 500 mA rcd at the mains position. A lack of discrimination may mean that essential services for milking or ventilation are shut down for trivial faults (see Fig. 10.3).

It is easy to understand that a local fuse with a small size wire will react more quickly to any fault than a heavy main fuse. Circuit-breakers have a similar characteristic built in. An rcd is different. The essential characteristic of a shock-protection rcd is that it shall react *instantaneously* to a fault. For practical purposes, instantaneous means up to 40 ms.

Therefore, reverting back to the example of a 30 mA protected socket in series with a 500 mA rcd, any fault in excess of 500 mA will cause both devices to operate instantaneously. Theoretically both are equally sensitive to this fault. In practice, age, working conditions and manufacturing tolerances will bring out one or other device first. But there is no planned discrimination.

The solution is to use a special time delay rcd for the upstream protection. The delay will be in terms of milliseconds but it should be sufficient to ensure that the local device normally operates first. Suppliers of rcds should be consulted.

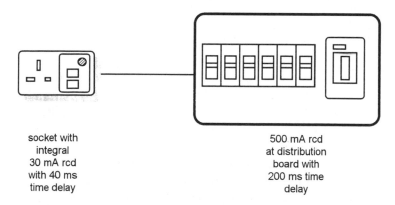

socket with
integral
30 mA rcd
with 40 ms
time delay

500 mA rcd
at distribution
board with
200 ms time
delay

Figure 10.3 **Time discrimination with series rcd protection.**

Shock protection

Earthing and bonding techniques are designed to give *indirect* shock protection, i.e. a shock derived from metalwork that is not live in normal use.

A *direct* shock is received when there is access to a live part when the primary protection of insulation is removed or damaged. 'Supplementary' protection is recommended for this condition and can be provided by an rcd with an operating current of 30 mA that will disconnect a 150 mA fault within 40 ms. Do not confuse supplementary bonding with supplementary protection.

605–08 ❑ Supplementary bonding is applied to create an equipotential zone and reduce indirect shock hazards under normal conditions.

❑ Supplementary protection is with the use of a 30 mA rcd to reduce direct
412–06 shock hazards caused by the failure of primary protection, usually insulation.

605–03 On a farm or horticultural establishment, *every* socket circuit must be protected by an rcd with the above characteristics. There is no exclusion for heavy current sockets or those for use within buildings.

It is wise to limit the number of sockets which will be protected by a single
531–02 residual current device and preferable to use one rcd or rcbo for each outgoing way on a distribution board, rather than protecting the whole board with one 30 mA device. This applies especially in damp situations where unavoidable leakage from many appliances will increase the sensitivity of the rcd and cause nuisance tripping (see Fig. 10.4).

Fire protection

Barns and store rooms on agricultural premises contain a great deal of flammable materials. Grain processing operations produce large quantities of
605–10 dust. These locations create a significant fire hazard which may be initiated by a minor electrical fault or earth leakage.

Regulations require that *every* part of a farm installation shall be protected by an rcd with a rating of up to 0.5 A (500 mA). This is not over-sensitive and nuisance tripping should not be a problem. The rating is sufficient to give an element of fire prevention. If it is imperative to reduce nuisance tripping as far as possible, it would be a good idea to subdivide the installation with a separate 200 mA rcd for each section.

Referring back to the sections on discrimination and shock protection it will be seen that this fire protection rcd will need to have time delay characteristics to ensure that initial response will arise from 30 mA devices elsewhere in the installation.

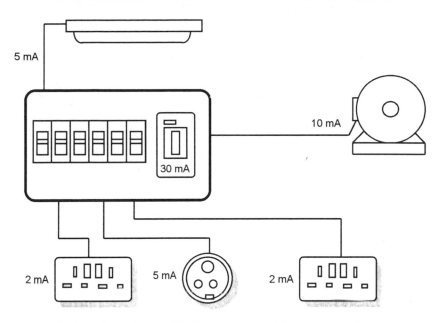

Figure 10.4 **Cumulative effect of normal leakages over-sensitizes 30 mA rcd.**

Omission of 500 mA rcd

Mention was made earlier of the importance of maintaining an electrical service to certain animal husbandry situations. Examples are where continuity of supply is required for milking and, even more essential, for animal life support ventilation in chicken and pig units. Unnecessary electrical failures must be avoided.

In these circumstances a decision may be made to omit the overall 500 mA rcd fire protection. This will be a judgement based on experience. Omission of a safety device should never be lightly undertaken. In the case of fire risks, insurance claims may be affected. The alternative may be to have a power failure alarm to ensure rapid response to clear the fault and restore power.

605–03   Regardless of the above consideration, local supplementary shock protection with a 30 mA rcd at sockets, must never be omitted.

Switchgear

512–06 Distribution boards and switchgear are frequently located in places with adverse environmental conditions. Wherever necessary, equipment must be specified to suit the conditions. This is a requirement of the Wiring Regulations and H & SE Electricity at Work Regulations. Standard industrial equipment may not be adequate and domestic type consumer units are usually quite unsuitable.

605–11   Equipment should be selected with the following conditions in mind:

❑ *Dust resistant (IP5X)*

522–04   Grain, chemicals, animal foodstuffs and woodcutting all produce vast quantities of dust.

❑ *Water resistant (IPX2 to IPX5)*

522–03   Rain, driving wind, condensation and hosing. The degree of protection must be specified. Non-metallic enclosures are always preferable.

❑ *Vermin resistant*

605–12   Rats, mice, birds and insects will all produce problems unless appropriate equipment is selected. Switchgear tends to be warm and inviting for nest building. Even when the gear is completely sealed, nests may be constructed behind enclosures.

❑ *Maintainable and user friendly*

341–01   Farm operatives may not be electrically qualified but speedy attention is often required to restore power.

Fuses are prone to abuse which can create hazardous fire conditions. Circuit-breakers are preferable for unskilled operation, but care must be taken to ensure that their enclosure satisfies protective requirements for dust and vermin, etc. An ordinary mcb or rcd is not dustproof and future efficiency could be restricted by a jammed mechanism. Frequent testing and operation helps to keep the mechanism free, but adequate dust protection is essential.

Wiring systems

Metal conduit and trunking gives the highest physical protection but the primary consideration under farm conditions is corrosion. Even galvanized
605–11   conduit rusts at unprotected joints. This can be overcome if suitable paint is applied but cuts in galvanized trunking are difficult to protect. The solution is to move away from metallic wiring systems.

Non-metallic wiring systems

Plastic conduit and trunking will resist all the usual agro-chemical materials and water. Fears about attack by vermin are usually unfounded unless conduit or trunking blocks holes which have previously been used as traffic routes by rodents. Rats and mice may sometimes eat soft polythene or *plasticized* PVC cable insulation but they are not normally interested in *unplasticized* PVC conduit and trunking.

The installation of plastic conduit and trunking is quick and simple. Care and attention must be given to workmanship:

❑ *Physical protection*
Conduit must not be run in places where it is susceptible to impact or otherwise abused.
As far as possible keep all circuit runs up high.
Adequate fixings are essential to eliminate sag.

❑ *Temperature*
Standard PVC conduit and trunking is suitable for most ambient conditions. It is extensively used in both cold stores and tropical climates where non-corrosive features are paramount.

Expansion will be a problem unless provision is made on long runs with the use of special expansion couplers as supplied by system manufacturers. Under UK conditions, between summer and winter extremes of temperatures, it may be anticipated that PVC will expand by approximately 1 mm per metre of run. This movement will mostly be taken up at bends but for straight runs, at least one expansion coupler should be fitted every 10 m. This, plus an adequate number of fixing saddles, will virtually eliminate sagging problems on exposed walls or under glass in hothouses. Saddles should permit longitudinal movement of conduit.

❑ *Water*
Generally speaking, neither plastic cable nor conduit suffers from the occasional wetting.

522 03   Complete sealing is almost impossible, therefore drainage is more important than sealing. Conduit and trunking should have drainage holes at appropriate places and conduit entries should be taken into the bottom of fittings. It will be noted that manufacturers of weather-resistant plastic accessories usually provide a small break-out drainage facility in the base of the mounting box.

❑ *Solar radiation*
522–11   Ultraviolet radiation from direct sunlight will biodegrade many unprotected plastics.

White conduit and trunkings should be avoided in sunny positions. The colouring additive in black PVC compound inhibits the penetration of ultraviolet light beyond a surface layer which may discolour.

Armoured cable

Steel wire armoured cables with PVC sheathing are suitable for most farming applications provided certain precautionary measures are taken.

❏ *Corrosion*
Special care is required to prevent corrosion of exposed armouring at joints.

❏ *Continuity*
It is doubtful whether permanent and reliable continuity can be obtained where swa is connected to metallic enclosures. It is always advisable to carry a separate cpc within the cable, or externally if a spare core is not available within the cable.

❏ *Water immersion*
The outer PVC sheathing of swa cable is good for most weather-resistant applications but may not be suitable for continuous immersion in water.

Where immersion is a requirement, the application should be discussed with the cable manufacturer who will supply an armoured cable with an outer sheathing which is submersible. An alternative is to install the PVC-sheathed swa cable in a continuous polythene pipe or waterproof duct.

Twin and earth cable
605–12

Ordinary PVC-insulated twin and earth cables are suitable for many farming applications where there is no need for further physical protection. Care should be taken in locations infested with rats or mice. Farm animals, horses and dogs have also been known to chew through insulation. Electrical systems should never be installed within reach of animals. Consideration should be given to possible changes of use of barns or temporary stocking arrangements in bad weather.

Overhead and underground wiring
522–08

It is often debatable which route to take. Much depends upon farm working conditions and future plans.

Overhead Wiring

Spans up to 3 metres

May be used in accordance with the following cable arrangements:

❏ PVC-sheathed and similar cables without intermediate support in positions inaccessible to vehicular traffic at a minimum height of 3.5 m.
❏ Cables as above but enclosed in a single span of 20 mm or larger, steel conduit in positions inaccessible to vehicular traffic at a minimum height of 3.0 m.

❑ Cables supported by a catenary wire, with support spacings as recommended by the manufacturer, in positions inaccessible to vehicular traffic at a minimum height of 3.5 m.

The minimum height of all cable spans above ground for positions accessible to vehicles is 5.2 m. Where there are road crossings the minimum height is increased to 5.8 m.

Underground cables
522–06

For very short routes between adjacent buildings, PVC-sheathed cable may be installed in buried conduit or pipes which are corrosion resistant taking into account the nature of the soil and contaminates present.

Most farm underground cables should be of PVC sheathed steel wire armoured. It is best if cables are enclosed in earthenware pipes but otherwise cable tiles or marking tape must be used.

The Regulations have no requirements for the minimum depth for burial of cables, except that it must be sufficient to avoid damage during normal use of the premises. This is always an on-site judgement and a suitable depth may be different on different parts of the farm.

General rules regarding farm electrical installations

This section summarizes some of the important Regulations for agricultural, horticultural and similar installations.

❑ Shock protection against direct contact is given with the use of suitable wiring enclosures installed in safe positions. Secondary protection is
605–03   achieved with the use of a 30 mA rcd.

❑ Shock protection against indirect contact is given with good earthing and
605–04   bonding practices.

❑ Wherever possible the use of all-insulated systems and enclosures is recommended. Class II all-insulated switchgear is available. This may be more expensive than industrial steel enclosures but maintenance costs
605–11   should be reduced.

❑ Wiring and switchgear must always be inaccessible to livestock. This applies to lighting switches in stable blocks in locations where horses may be temporarily tied up or in piggeries where barriers are moved about to
605–12   suit farrowing requirements.

❑ The positioning of emergency switching devices must take note of animal
605–13   behaviour, including panic.

605–10 ❐ Additional fire protection is given with the use of a 500 mA rcd.

❐ Heaters must be positioned to minimize any risk of burns to animals and fire. The consequences of fire must be considered in buildings where fodder and bedding is stored. Radiant heater elements should be at least 0.5 m from animals or combustible materials. This especially applies in farrowing pens, kennels and catteries.

600–01 ❐ Special Regulations apply to farm conditions but these are *additional* to standard Regulations and not alternatives.

Chapter 11
Isolation and Switching

This chapter deals extensively with safety requirements covered by the Regulations. Unless otherwise indicated, the reader must assume that statements are regulatory, i.e. there is no designer or installer choice on the subject.

If there is an element of judgement or choice in a design this will be made clear. Where such a decision has been made, the person signing the completion certificate as designer carries full responsibility. A design decision is acceptable provided that a competent judgement has been made in good faith, with safety the prime consideration.

Terminology

It must be emphasized that the use of correct terminology is essential for the understanding of technicalities. Words such as *switching* or *local isolation* are meaningless unless everyone is using them in the same way. If there is doubt about the interpretation of a word, it is always worth checking the exact definitions given in the Wiring Regulations.

Remember that both phase and neutral conductors are defined as live conductors.

Isolation and switching

There are four categories which should not be confused:

❏ *Functional switching*

537–05   This is the control for the user of equipment. The on-off facility. This may be the switch on a socket or the light switch on the wall. It may be the switch on an appliance or piece of machinery that also gives variable speeds or reversing control.

For some items of equipment where continuous operation is essential, e.g. a refrigerator, there may be no functional switch, or the functional switching is achieved by pulling out the plug.

461–01   ❏   *Isolation*
537–02   This means disconnecting the equipment or installation from the source of energy. The purpose of isolation is to enable work to be carried out by a skilled person on otherwise live parts. Pulling out the plug may be sufficient to isolate an appliance. As a universal rule, every installation must have means of isolation at the mains position.

462–01   ❏   *Switching off for mechanical maintenance*
537–03   This is intended for a non-technical operative or a cleaner to maintain the non-electrical parts of a machine, or for a householder to change a lamp.

    ❏   *Emergency switching*
463–01   This means disconnection from the supply to remove an unexpected
537–04   hazard. Not every installation or item of equipment requires emergency switching.

It can be seen that two or more of these switching applications can be
476–01   combined in one item of switchgear, or even a plug and socket. In other cases, equipment may not need switching for every purpose, for example an immersion heater does not require mechanical maintenance or emergency switching.

Isolation

Main switch

460–01   Every installation must start with a consumer's main switch. On very rare occasions this provision may be given by the supply company but permission must be obtained before omitting the consumer's switch.

    In most cases on TN systems, isolation is achieved by disconnection of all phase conductors with a suitable switching device. The one common exception is at the main switch on a single-phase installation. This must be double pole in order to switch phase and neutral.

    For a three-phase system the neutral need not be isolated. Any solid linkage within a three-phase isolator must be accessible only to a skilled person who must need to use a tool to remove the link.

461–01   The main switch for a TT or IT installation must isolate all live conductors. This means four-pole switching on a three-phase system.

    In some cases a building will have more than one electrical installation. For example, normal and off-peak services. In such cases the installations may be considered separately, each with its own main switch.

Security and lockability

Where an item of equipment contains live parts which may need attention by a skilled person, means of isolation must be provided. A British Standard specifies the exact requirements for an isolator which must have a positive mechanism to ensure that adequate contact is achieved.

461–01   One isolator may be used for several circuits or several items of equipment unless safety or convenience demands that separate isolation is essential.

Isolators must be installed in such a way as to prevent unintentional reclosure. Furthermore, an isolator being used for a particular function must be under the control of the person carrying out the electrical work. This may be achieved by having an isolator next to the motor or equipment being examined, or by having a secure lock off facility at a remote position.

476–02 ❏   Isolating switches usually have the facility for padlock attachment. It is not
537–02   sufficient to push a piece of wire through the padlock hole or hang up a notice.

❏   Special locking attachments are available for a standard mcb.
Alternatively the mcb board may be locked shut to prevent unauthorized resetting of a breaker. Care must be taken where there is more than one key that will open the distribution board.

❏   A local plug and socket arrangement provides admirable isolation provided that it is under the control of the operative requiring protection.

❏   Isolation by the removal of fuses is not normally acceptable unless there is no possibility of replacement or substitution by another person or a lockable security barrier is used.

Mechanical maintenance

Often the device used for isolation doubles up as the switch for mechanical
462–01   maintenance. Similar security rules apply.

537–03 ❏   In some circumstances the mechanical maintenance switch may be in the control circuit of equipment but if so it could not then double as an isolator. Where it is acceptable care must be taken to prevent automatic restarting or inadvertent operation by another person. For example, push button control is not appropriate without a lock off facility.

❏   The switch must be capable of breaking full load current. Therefore the removal of fuses is unacceptable.

❏   Account must be taken of the possibility that the person carrying out mechanical maintenance may not be electrically skilled.

❏   The location of the correct switch must be made obvious, either by its proximity to the task in hand or by labelling.

❏   The ON or OFF condition must be clearly indicated.

Emergency switching

The isolation device or switch for mechanical maintenance often also fulfils the function of an emergency switch. By its very nature an emergency stopping device must be able to operate on full load current and act as directly as possible on the supply.

476–03
537–04

❏ An emergency switch must be provided for any part of an installation where it may be necessary to prevent or remove danger.

❏ Where a risk of electric shock is involved, the rules for isolation of live conductors apply.

❏ A means of emergency stopping must be provided wherever electrically-driven machinery may give rise to danger from mechanical movement.

❏ The means of emergency switching or stopping must be durably marked, readily accessible and easily operated.

❏ An emergency stopping device may be incorporated in an appliance or as a part of the electrical installation. If separate from an appliance, one device may control more than one appliance, provided that it is readily accessible.

❏ Care must be taken to prevent unexpected restarting where the emergency stopping device is a push-button or if there is more than one means of restarting.

Additional special Regulations apply to the fireman's emergency switch for high voltage lighting installations and hazardous areas.

Examples of emergency switching

Emergency switching is necessary to protect people from danger which at the time of an incident may have been unexpected but should have been seen as a possibility at the time of design.

Domestic

Two domestic situations where emergency switching is required illustrate the point:

❏ A kitchen cooker fire is not usually anticipated by the user, but it is not an uncommon occurrence.

❏ A sink waste disposal unit can cause sudden physical injury due to careless use or interference with the works.

Both these situations require clearly identifiable and accessible emergency switching.

Other domestic appliances either have built-in emergency switching or do not have any foreseeable emergency condition:

❏ Hand-held motorized appliances usually have fingertip controls.

❏ Portable electric heaters have built-in switches or can be readily disconnected at the socket outlet.

❏ Appliances such as refrigerators or immersion heaters do not have a foreseeable emergency condition.

Industrial

In industrial and commercial type locations consideration must be given to the use of the equipment and possible hazards.

❏ Machine tools require positive emergency stopping. This is usually incorporated in a no-volt starter circuit.
❏ Continuous push-button pressure may be required for the operation of lifting equipment, or a 'dead-man's handle' principle applied.
❏ Electro-mechanical fail-safe braking may be essential to bring machinery to a speedy halt if a person becomes caught in rotating parts.
Automatic braking may also be required if the mains power fails to prevent uncontrolled movement of lifting machinery.
❏ Multiple emergency stopping facilities may be essential in training workshops, laboratories and kitchens.

Emergency stopping decisions

The above examples indicate that this is a wide and specialized subject. There is no single answer to *every* condition.

Wherever there is a non-standard need for the speedy disconnection of electrical power to equipment or a complete installation, the condition should be discussed with all interested parties and H & SE advisers.

Accessibility of switchgear

Two significant Regulations relate to fundamental requirements for safety:

130-06 ❏ Effective means, suitably placed ready for operation, shall be provided so that all voltage may be cut off from every installation, from *every* circuit thereof and from all equipment, as may be necessary to prevent or remove danger.

❏ Every piece of equipment which requires operation or attention by a
130–07 person shall be so installed that adequate and safe means of access and working space are afforded for such operation or attention.

These Regulations support similar statements in the H & SE Electricity at Work Regulations. Therefore in any working situation, they may be considered to be mandatory and may also apply to work in progress on domestic projects.

Switches for any purpose should always remain accessible. In some situations it may be decided to lock switchroom doors to prevent interference by children or unauthorized persons. In these circumstances the door should carry a notice indicating the presence of electrical switchgear and giving advice upon access. The key may be provided behind a break-glass facility or on a caretaker's keyboard. Naming an individual is useless and retaining the key within a locked drawer or room invalidates the requirement for ready operation.

The installation of domestic switchgear in kitchen cupboards should always be avoided. These inevitably become overcrowded with cooking utensils, shopping and cleaning materials.

Labelling and notices

514–11   Unless there is no possibility of confusion every item of switchgear should be labelled as to its function. If there are two or more main switches in a building, 476–01 each should be clearly marked to show which part of the system is being controlled. This may apply to normal and off-peak services or separate controls for dedicated computer supplies.

In an emergency it will be equally important for the fire brigade or an unskilled occupant to know what is being switched off and what is not being switched off. Sometimes there are two supplies into a building with isolation in different places. Details should be shown at all locations.

Notices and labelling must be permanent and clearly legible. Diagrams are often useful. Notices must be displayed near the switchgear. Installation manuals in the manager's office are of no value in an emergency.

Chapter 12

A Village Sports Centre

For reasons that will become obvious, this building is located on the edge of a sports playing field, remote from other buildings and on a rural electricity supply.

It could be a football pavilion, a golf club-house, or a health centre. The design could be combined with that for a small swimming pool to broaden the scope. Alternatively, the plans could be adapted for a school sports hall with a supply taken from the main building. Similar Regulations and conditions apply to any isolated building equipped with changing and shower facilities.

Special conditions

Because of the remoteness of this building there may be problems related to voltage gradients in the supply pme earth/neutral. The problem often becomes apparent when people feel electric shocks when using showers. The condition usually occurs in buildings with relatively heavy three-phase loads with an element of unbalance. Therefore, some off-peak heating has been added to this design, with all-electric hot water.

Note that this is a very generalized lighting and heating scheme and is intended only to illustrate electrical installation aspects.

Two Codes of Practice

❑   BS 5588 Part 6, *Fire Precautions for Places of Assembly.*
❑   Home Office Guide to Fire Precautions in Existing Places of Entertainment and Like Premises.

Although this is a small sports hall, it has been designed with facilities for social functions. This may involve dancing, entertainment and pop-groups. The Codes deal with all sizes of building and will probably be used by the architect or developer for design and construction.

This electrical design will mention recommendations from these Codes, both of which are advisory and not mandatory. However, it would be most unwise to ignore the recommendations which are likely to be applied by the

Local Authority for licensing purposes. The installer should obtain a reference copy of both documents.

Other interested parties

Most consultations will be through the builder, the client or the developer. Often schemes, such as this, are commissioned by a local charity with volunteer committee members directing the operations on an ad hoc basis. The importance of a full specification cannot be overstated.

The installation subcontractor must be certain that appropriate advice has been taken from all relevant authorities.

- ❐ *Electricity company*
  It is the developer's responsibility to negotiate for a supply. This job is usually delegated to the electrical installer. Make sure that a suitable electricity service can be taken to this remote building and check up on the cost. It has been known for this to be prohibitive.
- ❐ *Local licensing authority*
  This building will be used for social functions. Safety and environmental factors will be important. Perhaps a children's play group will use the premises.
- ❐ *Fire authority*
  Emergency lighting and fire alarms. These will be arranged in conjunction with door exits and fire-fighting facilities.
- ❐ *Client's insurers*
  Fire and personal safety requirements and the possibility of vandalism.
- ❐ *Beer supplier*
  For details of chilling and dispensing equipment.
- ❐ *Future users*
  If possible meet up with clubs and groups who will be using the facilities. There may, for example, be implications on the lighting design for badminton or volleyball activity.

Building details

An early study of the building construction is essential before commencing the electrical design (see Fig. 12.1 which also shows lighting and socket outlet provision).

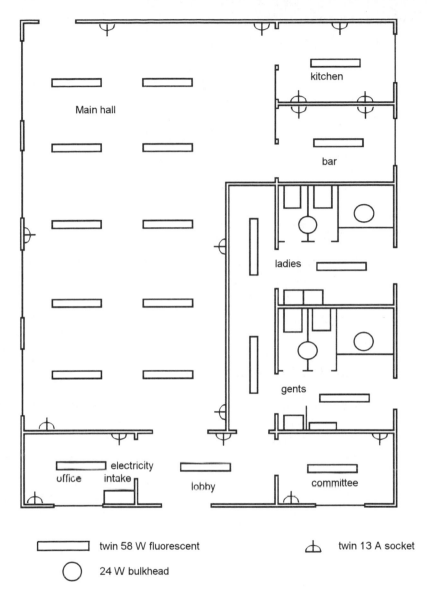

twin 58 W fluorescent ⌐⊥ twin 13 A socket

○ 24 W bulkhead

Figure 12.1 **Lighting and 13 A sockets in sports hall.**

Structure and finishes

This is a new project. The general construction is lightweight, economy being the top priority. The building is 300 m from the nearest electricity supply.

- Single storey, on flat ground.
- Total floor area: approximately 550 m^2.
- *Walls*
  External – cavity brick or building block.
  Internal – single building block.
  Wall finishes: decorated fair faced in hall and side rooms; not plastered; tiled in shower/toilet areas.
- *Roof*
  Pitched roof with timber trusses.
- *Ceilings*
  Plasterboard with good access to a roof space; glass fibre insulation above the ceiling.
- *Floor*
  Concrete base throughout with plastic floor tiles or ceramic tiles, as appropriate.

Electricity supply and requirements

After negotiation, the electricity company has agreed to provide an overhead service from the local pme network. This will enter the building in the entrance lobby, as shown on the layout.
    A good two-part tariff has been offered for an all-electric building.

- Space heating will be a mix of storage heaters and direct-acting heaters.
- Water heating will be with direct-acting showers and local sink water heaters. There is no stored hot water.

See Fig. 12.2 for details of the heating system.

Off peak tariff

Block storage heaters (15 × 3 kW) will be installed in the main hall and side rooms.

Normal tariff

- *Changing rooms*
  4 × 7 kW electric showers,
  2 × 3 kW quartz radiant heaters,
  2 × 2.5 kW sink heaters,
  20 × 0.18 kW tubular heaters.

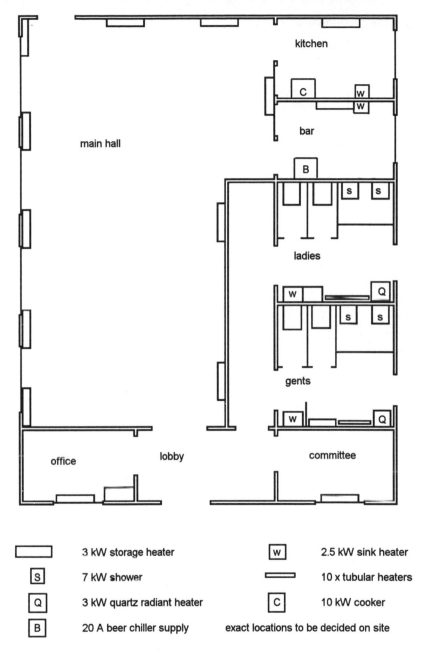

Figure 12.2 **Heating loads in sports hall.**

❏ *Kitchen and bar*
2 × 2.5 kW sink heaters,
10 kW domestic type cooker,
20 A single-phase supply for beer chiller, etc.
❏ *Socket outlets*
16 × twin 13 A sockets.
❏ *Lighting*
See Fig. 12.1 for details of the lighting scheme.
19 × twin 58 W fluorescent fittings,
4 × 24 W fluorescent bulkheads.

Load assessment and diversity

311–01   It is important to calculate maximum demand characteristics at an early stage. This will affect much of the electrical design and will be required to confirm that the electricity company can make a supply available.

The IEE Guidance Note provides no diversity comparisons for this type of building. The designer must consult the client and make judgements upon electrical utilization. Conclusions reached by the application of diversity figures only apply to the estimated total maximum demand and not to the sizing of final circuits and cables. This is not an accurate study and many figures have been rounded off.

Off-peak heating

There is no diversity over the storage heaters which will all operate simultaneously whilst the rest of the building is in use.

$$15 \times 3 \text{ kW heaters} \quad = \quad 45 \text{ kW}$$

This will be spread across three phases   =   62.5 A per phase.

Normal tariff

Showers

All four may operate simultaneously for relatively short periods of time, usually when off-peak heating is off. An allowance of 75% has been given.

$$4 \times 7 \text{ kW} \times 75\% \quad = \quad 87.5 \text{ A}$$

Changing room

All quartz heaters and tubular heaters will be used for extended periods of time. The tubular heaters will be thermostatically controlled but 100% heating will inevitably be required at crucial times. No diversity can be allowed.

$$2 \times 3 \text{ kW quartz heaters} \quad = \quad 25 \text{ A}$$
$$20 \times 0.18 \text{ kW tubular heaters} \quad = \quad 15 \text{ A}$$

Sink water heaters

These will only be used for short periods of time. Allow for 25% diversity.

$$4 \times 2.5 \text{ kW} \times 25\% \quad = \quad 10.4 \text{ A}$$

Cooker

Only occasional use at normal times for short periods of time. Allow 25% diversity.

$$10 \text{ kW} \times 25\% \quad = \quad 10.4 \text{ A}$$

Socket outlets

Probably these will be divided across three circuits. The greatest potential load will be in the kitchen. In other locations there may be occasional 'emergency' use of portable heaters. Otherwise loads will be small.

$$\text{Assume 30 A max connected load} \quad = \quad 30 \text{ A}$$

Beer chiller

This will be an intermittent load. Allow 50% diversity.

$$20 \text{ A} \times 50\% \quad = \quad 10 \text{ A}$$

Lighting

All lighting is fluorescent, therefore a 1.8 factor is applied to lamp ratings to allow for control gear losses. On occasions lighting will be fully used for extended periods. A 90% allowance is given.

$$19 \times 2 \times 58 \times 1.8 \times 90\% \quad = \quad 3.6 \text{ kW plus}$$
$$4 \times 24 \times 1.8 \times 90\% \quad = \quad 0.2 \text{ kW}$$
$$= \quad 16.0 \text{ A}$$

Total estimated maximum current demand

$$\text{Off peak} \quad = \quad 187.0 \text{ A or}$$
$$62.3 \text{ A per phase}$$
$$\text{Normal} \quad = \quad 218 \text{ A or}$$
$$72.6 \text{ A per phase}$$

The electricity company should be given a figure of 135 A per phase on the assumption that load balancing is reasonable. Ultimately the supply company will use its own experience to determine the service provision.

The designer must consider these loads for switchgear capacity and circuitry, making due allowance for unbalanced phases.

Figure 12.3 shows a suggested project specification.

Wiring systems

At this point note is taken of the Codes of Practice issued by the British Standards Institution and the Home Office. These specify that all wiring should be enclosed by suitable protection against physical damage. Acceptable systems for mains wiring are steel wire armoured or mineral insulated cables and steel or PVC conduit. Unprotected sheathed cables may only be used for extra-low-voltage non-emergency circuits.

512–06 On this project the obvious main wiring route is through the ceiling/roof space. Cables in this location could be thought of as not subject to physical maltreatment and there is therefore a choice of systems.

Insulated and sheathed cables

With economy in mind thoughts will initially run to the use of twin and earth cables. Apart from the Codes of Practice, there are practical reasons why sheathed cables are unsuited to the building.

Taking into account the Wiring Regulations' requirement for good workmanship, the use of unenclosed wiring is not recommended anywhere where there are large groups of cables. Consider initially the number of circuits. Unenclosed cables would need to be untangled before extensive clipping can be applied. One solution to the problem of cable fixing is to carry groups of sheathed cables through trunking which has to be very large to handle the bulk without overcrowding.

Surface run cable drops down wall to heaters and sockets must be enclosed in conduit or mini-trunking for both protective and appearance reasons. The scheme eventually becomes a mixture of trunking, conduit and mini-trunking with gaps between where sheathed cables are exposed.

Steel conduit and trunking

It is generally best not to mix steel and PVC systems. Complications arise with earthing and jointing. Steel conduit and trunking would undoubtedly make a good job, but the installation costs could hardly be justified.

If the large room was used as a school gymnasium and some damage to wiring could be anticipated, then perhaps steel conduit would be desirable. However, it is a golden rule that the best method of protecting electrical cables is to locate them along safe routes.

Project Specification BS 7671
Sports Centre

Name....................... Location.......................
Reference................ Date.....................

3 - phase 240 V 50 Hz. TN-C-S. Supply fuse 160 A BS 1361 or BS 88

PFC less than 16 kA. Earth loop impedance less than 0.35
Normal supply 125 A TP Main switch Off peak supply 125 A Main switch
2 Type B Distribution boards

Circuits	Rating (A)	Cable size mm^2	Max length m	Lights/points g = gang
Normal supply distribution board 125 A isolator 18 ways				
lights 3 circuits	10	1.5	55	23
13 A sockets 3 circuits	32	2.5	66	16 x 2g
showers 4 circuits	32	6.0	40	4
quartz heaters 2 circuits	16	2.5	30	2
tubular heaters 2 circuits	10	1.5	30	20
sink heaters 4 circuits	16	2.5	30	4
cooker	32	6.0	40	1
beer chiller	20	2.5	30	1
Off peak distribution board 125 A isolator 18 ways				
Storage heaters 15 circuits	16	2.5	30	15

Figure 12.3   **Project specification for village sports centre.**

Emergency systems

Details of fire alarm and emergency lighting provision should be added to the specification.

Plastic conduit and trunking

This is to be the preferred system for appearance, convenience and economy. Material costs may be higher than with sheathed cables but the use of single-core cables, smaller trunking, and speed of installation will compensate for this.

A recommendation here is that conduit and trunking should be obtained from the same manufacturer. This ensures compatibility with appearance and fit.

At first sight, the most practical idea seems to be to run a large trunking above the ceiling, through the centre of the hall, and tee-off with conduits to outlet positions. This may cause problems when cable sizing is considered when applying derating factors for groups. Note that the storage heater cables App. 4 will be fully loaded simultaneously. Excessive grouping may require larger than normal cables.

A better approach will be to take two or three smaller lateral trunking runs in the roof space along the length of the hall. Each of these runs would contain a mix of lighting, power and heater cables to local outlet positions. This reduces the need for derating-grouped, heavily-loaded cables. Single-core insulated cables will be installed with complete trunking enclosure and conduit drops down the walls. Common circuit protective conductors may be used as described in the section on earthing.

An alternative which may be considered, for reasons of appearance, is to 521–07 enclose wall cable drops in mini-trunking. A problem then arises with bends in the roof space at the top of the wall and connections into the main trunking. Even the best mini-trunking systems do not have the same easy bending and jointing facility as with PVC conduit.

One solution to the above situation is to come away from the trunking horizontally with plastic conduit and convert at the wall/ceiling junction with a blanked-off switchbox having conduit entry on one side and mini-trunking on the other.

Circuitry and cable sizing

Grouping factors must be considered to determine cable sizes but firstly it is App. 4 worth looking at the circuits and establishing 'ungrouped' sizes (see Table 12.1).

Cable grouping factors

The Regulations give derating factors for groups of cables of all one size and apply the factors to fully loaded conditions. This situation rarely applies, so some sort of judgement must be made.

Table 12.1 **Circuits showing full load requirements.**

Circuit	Full load (A)	mcb (A)	cable size, ungrouped (mm^2)	length of run, max (m)
Lights (3 circuits)	6	10	1.0	35
or	6	10	1.5	55
Sockets (3 circuits)	Ring	30	2.5	70
Storage heaters (15 circuits)	12.5	16	2.5	30
Showers (4 circuits)	29	32	6.0	40
Quartz heaters (2 circuits)	12.5	16	2.5	30
Tubular heaters (2 circuits)	7.5	10	1.5	55
Sink water heaters (4 circuits)	10.4	16	2.5	30
Cooker	30	32	6.0	40
Beer chiller	20	20	2.5	30

In this project the only group of similar-sized, equally-loaded cables are the storage heater circuits. These are actually loaded at less than half the tabulated current rating. Under such conditions, provided that thermal insulation is avoided, derating of heater cables may be ignored.

Using similar logic, it will be seen that all other circuit cables are sized conservatively in terms of current-carrying capacity. The reasoning is that voltage drop considerations are more significant. This is not an uncommon situation, and therefore it is suggested that the sequence of cable-sizing calculations should always start with voltage drop.

523–04   Thermal insulation is a consideration. There is glass fibre above the plasterboard ceiling. A sensible idea is to span trunking and conduit across the top of ceiling rafters, clear of thermal insulation. Care must be taken to ensure that adequate support is given to plastic conduit. It may be advisable to use timber bearers across spans in excess of about 2 m.

Phase	Circuits	Amps
	Table 12.2 **Phase balancing using diversified loads.**	
Red	5 storage heaters	62
	No. 1 sockets	10
	No. 2 sockets	10
	No. 3 sockets	10
	2 showers	44
	Total (10 circuits)	136
Yellow	5 storage heaters	62
	No. 1 lights	6
	2 showers	44
	4 sink water heaters	12
	Cooker	10
	Total (13 circuits)	134
Blue	5 storage heaters	62
	No. 2 lights	6
	No. 3 lights	6
	2 quartz heaters	25
	2 tubular heater circuits	15
	Beer chiller	20
	Total (12 circuits)	134

Arrangement of circuits

There is no Regulation which requires 415 V circuitry or equipment to be segregated. The only obligation is for suitable labelling. It is, however, good 514–10 practice to try to keep phases separate whenever possible. This consideration may be overruled when phase balancing is worked out (see Table 12.2). This schedule uses approximate, diversified current demands and produces a good balance.

Switchgear

There will be two three-phase Type B distribution boards.

❏ *Off peak*
    18 ways (15 used)   125 A main switch (four pole for TT)

❑ *Normal supply*
24 ways (20 used)   125 A main switch (four pole for TT)
Spare circuits may be used for emergency systems.

Shock protection

Socket outlets which may be used for entertainment purposes require 30 mA rcd protection to satisfy Home Office requirements. In the circumstances it will be wise to protect all socket outlet circuits at the distribution board. A 24-way distribution board will be adequate if each single way uses an rcbo.
412–06 The alternative of a combined rcd/mcb will each take up two ways, requiring a 30-way distribution board.

There is no requirement for an rcd elsewhere on the system with a pme supply. The requirements on a TT installation will be shown later in this chapter.

Earthing

A main earthing terminal will be established adjacent to the distribution boards. It is not acceptable to make use of one of the earthing bars within one distribution board and link into the other. An accessible and independent bar
542–04 will be required to terminate the electricity company's pme earthing conductor or the earthing conductor from a TT electrode. A main bond to the water supply will connect at this position and two circuit protective conductors will connect to the distribution boards.

Bonding

547–02 Standard bonding requirements apply for a 10 mm^2 connection to the main water intake position, but this size should be confirmed with the supply company.

An occasional problem

At the start of this chapter an indication was given that problems have been noted with pme systems supplying a remote building of this type. The condition is illustrated with reference to Fig. 12.4.

The supply company uses a combined earth/neutral conductor which in this case is shown as an overhead line. This conductor is connected to an earth electrode at the foot of the pole. The service divides at the meter cut-out, giving a neutral connection to the distribution board and an earth to the main

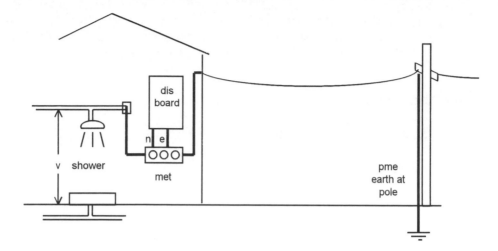

Figure 12.4 **Problem caused by volt drop through pme neutral.**

earthing terminal (met). Within the installation, a main bond has been taken to the water stop cock which takes the mains water through a plastic pipe.

All the plumbing in the building is now equipotential, with the supply earth/ neutral and all exposed conductive parts on storage heaters, etc. This creates standard pme conditions but with two significant features:

❑ Unless the load is perfectly balanced, the overhead neutral conductor will be carrying some load current. This results in volt drop and the 'earth' at the cut-out is at a slightly different potential to the local ground earth.
❑ Showers have a common tray with a direct water outflow into the ground. A person standing in the shower may notice a potential difference of perhaps 10 V between the shower head and the ground. This is dis-comforting and causes concern.

Solutions

❑ At the time of building construction a metallic grid should be put in the concrete base beneath the shower trays. This grid is then given a bonding connection across to the plumbing or the met. A full equipotential con-dition is thus created.
❑ With an existing building the provision of an earthing grid may not be practicable. A solution may be found by driving in one or more earth electrodes, outside the building and as near as possible to the shower drainage. Again, bonding connections should be made to the plumbing or met.
❑ If the problem is noticed on an existing installation neither of the above solutions may be practicable. The only solution may be to disconnect the

542–01   supply earth at the met and consider the installation to have a TT supply. This will necessitate obtaining an installation earth from an earth
413–02   electrode and having a three-phase rcd at the main switch position.

Requirements for a TT installation

There are special requirements for a TT installation. Guidance will be found in Chapter 7, Earthing and Bonding, and the following factors should be noted:

471–08   ❑   All socket outlets must be rcd protected. This condition has already been achieved in the design.

❑   The installation earth facility for shock protection will be with an earthing rod associated with an rcd at the mains position. This rcd should not be over-sensitive: a 200 mA or 500 mA rating is suggested. The corresponding maximum permissible resistance from any point in the installation through the electrode is 250 ohms and 100 ohms, respectively. However, any earth electrode test reading of more than 50 ohms must be considered to be unreliable and should be improved by installing deeper or additional rods.

531–02   ❑   To ensure discrimination in the event of a fault on a circuit with 30 mA protection, the main rcd should have time delay characteristics. This feature was covered in Chapter 10.

Chapter 13

An Indoor Swimming Pool

This is a domestic pool for private use and attached to a house. In general, similar rules apply for larger installations. These may be complicated by public access requirements and other sports facilities. Only actual swimming pool requirements are studied. The proposed design could be coupled with the sports centre in Chapter 12 and provision made for the pool's public use.

This study does not include designs for the electrical control of water treatment and pumping equipment which will be installed by the specialist pool installer.

Special conditions

A swimming pool is obviously a hazardous area for the utilization of electrical equipment. Extreme care must be taken with arrangements for electrical supplies to pool equipment and the installation in the pool area. This chapter
Sec. 602 can only advise on requirements covered by the special sections of the IEE Wiring Regulations devoted to the subjects of swimming pools and saunas. Potential installers should be familiar with these requirements.

Problems related to the appearance of inconvenient but generally harmless, non-equipotential pme conditions will be mentioned here with regard to swimming pool facilities. The subject is given extensive consideration for sports hall showers at the end of Chapter 12. Similar solutions would apply for any swimming pool situation.

Only a passing reference is made to an emergency lighting facility. The subject of security, emergency and telephone services should be discussed with the client, especially if usage extends beyond family domestic activities.
512–06 A major consideration is the highly corrosive and humid atmosphere. Special care must be taken in the selection of appropriate electrical equipment.

The Wiring Regulations specify zones around swimming pool areas where restrictions apply to the installation of electrical equipment. Suggestions are given for the application of these zones to a typical situation.

Emergency lighting

There is no requirement for emergency lighting with this private swimming pool, but the subject should be considered.

The electricity supply may fail on a dark evening. This would cause hazardous conditions for swimmers, especially children. It is recommended that at least two self-contained emergency luminaires should be provided, one at each end of the pool.

Other interested parties

Most consultations will be through the client, an architect or the builder. It is important that a full specification should be prepared in advance of tendering. Some aspects of electrical design affect the pool construction and should be considered before construction starts. Layout and decorative treatments need to take account of permissible lighting arrangements.

The electrical installer must be certain that appropriate advice has been taken from all relevant authorities.

☐ *Specialist pool installer*
  Electrical requirements for all pool equipment and servicing facilities.
☐ *Electricity company*
  It is the consumer's responsibility to ensure that an adequate supply is available. This job is usually delegated to the electrical installer. The local electricity company may have restrictions regarding the provision of a pme service.
☐ *Local authority*
  If there is any possibility that the pool will be used for public events such as a garden party or fete, the local authority's environmental health department should be approached on the subject of safety services.
☐ *Client's insurers*
  Are there any special fire and personal safety requirements?

Building details

Construction

☐ *Design*
  A new single storey purpose built structure to be constructed on private land adjoining a large house. Total floor area: approximately 110 m².
☐ *Floor*
  Concrete, tiled throughout.
☐ *Walls*
  Brick, fair faced externally and internally. Large glazed areas.

❏ *Roof*
Timber frame, lined underside with pinewood ceiling. The ceiling over the pool area is 3 m above floor level. The ceiling over the projecting leisure verandah is 2.25 m above floor level.

❏ *Changing area*
Walls: building block, tiled throughout.

❏ *Space and pool water heating*
Gas.

Electrical requirements

Full details of electrical loads are given in Table 13.1 and the electrical layout is shown in Fig. 13.1. The lighting system as shown is for load assessment 522–05 only. Manufacturers should be consulted regarding the suitability of products for this corrosive situation.

Table 13.1 Schedule of equipment.	
All single-phase equipment	
Location	*Single-phase equipment*
Lighting Pool area Verandah Plant room Changing room	5 × 150 W SON 5 × 100 W recessed downlighters 1 × 58 W fluorescent 4 × 24 W bulkhead fluorescent
De-humidifiers	2 × 20 A supplies
Plant Room Power	1 × 30 A supply
Socket outlets Verandah Pool services Plant room	1 × twin 13 A 2 × 16 A BS 4343 1 × twin 13 A 1 × 16 A BS 4343
Provision for drier Changing room	1 × 2 kW outlet

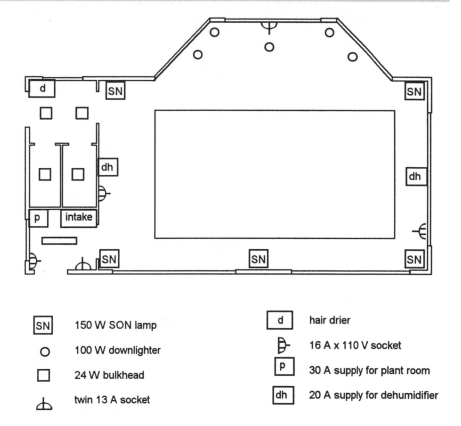

Figure 13.1 **Electricity provision in swimming pool.**

A suggested project specification is given in Fig. 13.2.

Zones

602–02 The area surrounding the pool which is in general use may become wet and must be zoned in accordance with standard requirements (see Fig. 13.3).

Zone A

This is the water volume within the pool basin including the above-water area up to deck level and any accessible apertures or ledges.

602–04 ❏ Only 12 V maximum SELV (safety, extra low voltage) equipment is permissible. This must be supplied from control gear outside Zones A and B.

This will allow only for protection by IPX8 (submersible) purpose-made

Project Specification BS 7671
Domestic Swimmng Pool

Name...................... Location........................
Reference................. Date......................

Single phase 240 V 50 Hz. TN-C-S fed from adjacent house Supply fuse 63 A BS 88

PFC less than 16 kA. Earth loop impedance less than 0.35 Consumer unit BS 5486 Four type 2 mcb circuits One type 3 mcb circuit Two 30 mA rcbo circuits

Circuits	mcb type	Rating (A)	Cable size (mm^2)	Max length (m)	Lights/points g = gang
1. Lights -SON	3	10	1.5	55	5
2. Lights	2	10	1.5	55	10
3. Sockets 13 A	rcbo	20	2.5	26	2 x 2 g
4. Sockets 16 A	rcbo	20	2.5	26	3
5. Dehumidifier 1	2	20	2.5	26	1
6. Dehumidifier 2	2	20	2.5	26	1
7. Plant room	2	32	6.0	40	1

Figure 13.2   **Project specification for swimming pool.**

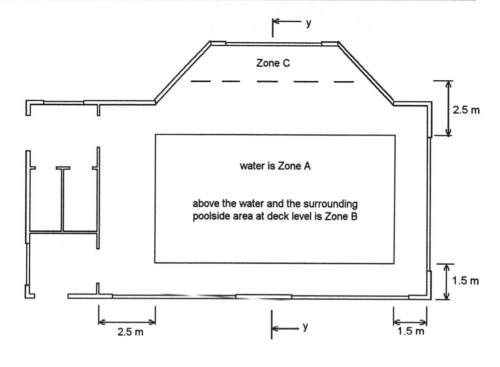

Plan view

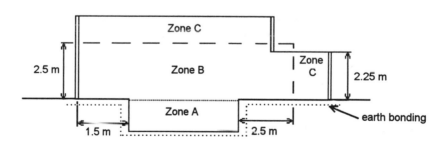

Cross section at y-y on plan

Figure 13.3 **Assessment of zones.**

underwater floodlights, each of which must be supplied from its own transformer with an open-circuit voltage not exceeding 18 V.

❏ The only wiring allowed is for equipment within the zone.
❏ No junction boxes switchgear or accessories are permissible.

Zone B

This is the area above Zone A, extending vertically to an arm's reach height of 602–05 2.5 m and full arm span of 2 m horizontally from the poolside. The 2.5 m overhead clearance applies above any steps or poolside raised areas including fountains or diving facilities.

❏ Only 12 V maximum SELV equipment is permissible, as in Zone A, except that IPX5 (hoseproof) minimum protection is adequate.
❏ The only wiring allowed is for equipment within Zones A and B.
❏ No junction boxes, switchgear or accessories are permissible except: If a socket outlet is essential and cannot be located outside Zone B, e.g. for a mechanized pool cover, a BS 4343 'Commando' type socket may be installed not less than 1.25 m from the pool edge and at least 0.3 m above the floor. This socket must have 30 mA rcd protection.

Zone C

This is the reach-out area extending horizontally 1.5 m from Zone B to a similar height of 2.5 m.

❏ All equipment must have IPX5 (hoseproof) protection.
❏ A BS 4343 'Commando' type socket with 30 mA rcd protection may be installed.
❏ Standard SELV equipment may be used.

Application of zoning to this project

Lighting

602–04 Lights on the ceiling above the pool would be more than 2.5 m high and outside any zone; however, this location should be avoided because of access for changing lamps. A better arrangement would be to mount lights at high level on the wall.

Luminaires must be out of reach, above 2.5 m minimum height within Zone B. Wall brackets at a lower level are not acceptable except where the poolside deck extends horizontally beyond 2.0 m.

On the project in question, the first 0.5 m of the low ceiling in the verandah area comes within Zone B and must be avoided for ceiling-mounted luminaires.

Lighting switches

602–07 Switchgear may be installed in Zone C but to avoid long-term corrosion problems a better location is in the ventilated changing area.

Sockets

602–08 The Regulations require that all sockets must be of the BS 4343 'Commando' style, with 30 mA protection and that they must be located more than 2 m from the poolside. This, therefore, limits their location to one end wall or the verandah area. The sockets shown are for use with pool cleaning equipment.

At the insistence of the customer, one 13 A 30 mA rcd-protected socket is required for a radio and tape player. This has been positioned on the farthest wall, at the extremity of Zone C. It probably contravenes the Wiring Regulations and a note to that effect will be added to the completion certificate. Extreme care must be taken in accepting any deviation from the regulations.

Dehumidifiers

These may be mounted as fixed appliances in Zone B or C areas provided that they are of a type specifically intended for use in swimming pool areas. Protection on a 30 mA rcd circuit is essential with a wired connection to the supply, not a plug and socket.

Changing Room/Shower Area

Swimming pool zoning does not apply in these locations. Regulations applicable to showers will apply. These include:

601–10 ❑ 13 A socket outlets are not permitted; neither is provision for the connection of portable appliances. If a hair drier is required this should be of the wall mounted type with a hot air nozzle or flexible hose.
BS 3535 shaver units are permitted.

601–08 ❑ Lighting switches must be out of reach of a person using the shower, or pull-cord operated.

601–12 ❑ Heaters with exposed elements must be out of reach of a person using the shower.

Loading and diversity

The Wiring Regulations offer no guidance on this type of installation.

Lighting

It is probable that all lights will be used together. No diversity will be allowed. For current loading calculations, discharge lamp ratings should be multiplied by 1.8 to take into account control gear losses.

$$\text{Pool SOL lighting} \quad \frac{5 \times 150 \times 1.8}{240} \quad = \quad 5.6 \text{ A}$$

$$\text{Verandah spotlights} \quad \frac{5 \times 100}{240} \quad = \quad 2.1 \text{ A}$$

$$\text{Changing room} \quad \frac{4 \times 24 \times 1.8}{240} \quad = \quad 0.7 \text{ A}$$

$$\text{Plant room} \quad \frac{1 \times 58 \times 1.8}{240} \quad = \quad 0.4 \text{ A}$$

$$\text{Total lighting load} \quad = \quad 8.8 \text{ A}$$

Dehumidifiers

These do not work continuously and the 20 A requirement is probably excessive. Assume 75% diversity.

$$2 \times 20 \times 75\% \quad = \quad 30 \text{ A}$$

Socket outlets and hair drier

There is no heating load. All sockets are for occasional use.

Assume 30 A maximum load at 33% diversity   =   20 A

Total maximum demand   =   59 A

A 60 A single-phase supply will be adequate.

Wiring systems

Corrosion is the major consideration. The Wiring Regulations specify that a
602–06 surface wiring system shall not employ the use of metallic conduit or trunking, or an exposed metallic cable sheath.

❐   Steel conduit and trunking is excluded completely.
❐   Mineral insulated cable (mics) with a continuous PVC sheathing may be acceptable provided that exposed copper or brass is protected at terminations.

Severe corrosion may be encountered where mics enters steel enclosures. For this reason it has been ruled out for this project.

❑ Steel wire armoured, PVC-sheathed cable would be acceptable provided that terminations could be protected against corrosion. In this instance the only large load that would warrant the use of this cable is the short run in the plant room for pool equipment.

❑ Twin and earth sheathed cables would be appropriate but in some areas surface wiring would require additional non-metallic cover for protection and appearance reasons. If sheathed cables are buried beneath plasterwork in these damp conditions, plastic capping should be used.

❑ Plastic conduit and trunking systems are ideally suited to these conditions and will be the chosen materials. Wherever possible non-metallic enclosures should be used.

Cable sizes

Provided that thermal insulation can be avoided, there are no special limitations requiring derating factors to be applied. Cables installed in plastic conduit or trunking will not be heavily loaded.

There may be a large group of mixed size cables in trunking near the distribution board. The 32 A supply for the plant room should be kept separate. This will be the case if swa cable is used for this short run.

Cable sizes are shown in Fig. 13.4.

Distribution board

A single-phase eight-way distribution board is adequate and this should have a non-metallic enclosure. Two lighting circuits are shown to avoid inconvenience in the event of the failure of one. SON discharge lamps do take a surge on start-up and type 3 or C circuit-breakers are advisable.

314–01

The 13 A sockets and the outlet for a hair drier may be run from one single module 20 A combined mcb/rcd (rcbo). Similar rcd protection is required for the BS 4343 sockets and a 16 A rating rcbo will give adequate protection. Although perhaps not strictly essential, rcd protection may be given to the dehumidifiers that are located in potentially wet areas.

602–07

It should be noted that an rcbo is a single pole device and the neutral remains connected after fault disconnection. It may be considered desirable to have double pole rcd protection. This will require a double mcb space for each unit in the distribution board.

Apart from on the SON lighting circuit all other circuit-breakers should be type 2, and unless the electricity company insists on a TT supply, no other rcd

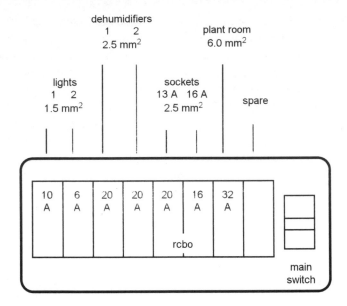

Figure 13.4 **Distribution board and cable sizes.**

protection is necessary. It is not advisable to give overall sensitive rcd protection in this situation where loss of lighting could be hazardous to pool users.

Isolation

The facility for isolation of all circuits will be at the distribution board. The main switch must have a lock-off facility and it is suggested that at least one suitable mcb lock should be provided for future use at the distribution board.

476–02

110 V supply

A 1 kVA 240/110 V transformer will be located in the plant room adjacent to the distribution board. It is important to check that the 110 V winding has a mid-point earth connection. There is no requirement for switching at BS 4343 sockets. This subject should be discussed with the client. It may be considered that switched sockets would provide speedy disconnection in an emergency situation. However, it may not always be safer to have a switch, which may encourage a user of cleaning equipment or a motorized pool cover to leave a plug permanently inserted. It is obviously safer to remove the plug for equipment not in use. Even reduced low voltage at 55 V to earth can be lethal in these wet surroundings.

Earthing

This subject must be carefully considered. All conventional earthing and main bonding should be carried out in accessible locations where connections may be separated for periodic testing. Earthing clamps to piping should be of the non-corrosive type, usually colour coded blue. Those coded red are not suitable for damp environments. Cable terminations at earth clamps should be crimped to ensure reliability throughout the life of the installation.

543–03

Local supplementary bonding

602–03 This is one location where the Wiring Regulations seek local supplementary bonding. Apart from using the word local there is no regulatory guidance on how this should be achieved. There is no requirement for a special supplementary bonding cable to be taken back to the distribution board, only that exposed conductive parts and extraneous conductive parts should be interconnected.

The definition of extraneous conductive parts in Chapter 7 mentions parts that may introduce a potential into an otherwise equipotential zone. It is clear that wall, floor mounted or removable handrails only have the local potential and therefore do not need bonding. Metalwork entering from outside the building should be cross-bonded to the local system.

In the project under examination it is suggested that the dehumidifiers should have cross-bonding to other exposed conductive parts of electrical equipment and earthy metalwork. This would include all structural steel or metallic plumbing within the zone.

Floor grid

The Wiring Regulations require that an equipotential floor grid be provided in all Zone B and C areas. This is obviously a point that needs discussion with the builder at an early date in the contract. This grid must be cross-bonded to the equipotential bonding system by means of accessible and reliable connections. Where there is more than one earthing grid the subject of interconnection should be carefully planned with access covers over sunken inspection traps. Detailed suggestions for this floor grid will be found in the IEE Guidance Note No 5.

Chapter 14
Cables and Wiring Systems

The choice of wiring systems for specific projects is increasing with developments in materials. There is currently great interest in the performance of cables in fires.

An electrical installation designer must take account of the range of conditions under which a cable will be used. Wiring Regulations list these under the heading of external influences.

External influences

App. 5

❐ *Environmental*
This relates to climatic influences whether caused naturally by weather and geographical features, or by man-made conditions. Different considerations would apply, for example in a refrigerated cold store and an abattoir on the same premises.

❐ *Utilization*
The occupancy of a building is important. A hospital for mentally handicapped patients will have completely different utilization features to a factory workshop.
The actual electrical energy requirement at the point of use often dictates the wiring system.
Safety features such as smoke production in a fire will be involved.

❐ *Building*
The construction and shape of a building will affect the routeing and protection required by the wiring system. The subject of fire resistance may also be significant.

Cost considerations

An installer will aim to select a wiring system which gives adequate technical features, at the best price. There is no merit in over-specification. Traditionally, emphasis has been given to the physical protection of wiring. Heavy gauge steel conduit and trunking is still used in many unsuitable locations where cheaper alternatives are aesthetically more appropriate and no less safe.

Labour costs are one of the most important factors in the economy of any project. By its nature, the installation of a wiring system is labour intensive.

The cost per metre of a cable is not the sole criterion when designing an installation.

Choosing suitable cable routes

Wiring may be physically protected either by enclosure within a robust outer covering or by selecting a suitable route which avoids the possibility of damage. The latter choice is always the most sensible and may provide the opportunity to use a lower cost system with less complex mechanical protection. It is therefore important to have details of the building construction before deciding upon the type of cable to use.

App. 4   Wherever possible a cable route should be chosen that avoids hot, damp or dusty atmospheres. Many of these detrimental conditions require derating factors to be used for cable sizing with consequential financial penalties.

Is armouring always necessary?

A typical example of over- engineering is with the almost universal use of steel wire armoured cables for sub-main distribution in commercial and public buildings. Usually these heavy cables are routed on cable tray through vertical or horizontal service ducts. There is no conceivable reason for armouring which is usually supplemented by a green/yellow circuit protective conductor either as an additional core or with a separate conductor. The result is that complicated continuity arrangements are required at glands for the sole purpose of earthing the superfluous armouring.

Non-armoured cables are available which are lighter and easier to install.

Fire barriers

527–01   In order to prevent the spread of fire and smoke, buildings are usually divided into zones, or compartments with appropriate fire resistant elements such as walls or floors. The effectiveness of this compartmentalization may become a matter of life and death during emergency evacuation of a building.

Holes through fire barriers

527–02   Oversize holes cut into the building structure will invalidate fire prevention measures. Open vertical cable ducts are notorious for producing a chimney effect which not only transmits smoke and fumes but may also provide sufficient oxygen to fuel a cable fire.

The Wiring Regulations require that where a wiring system passes through floors, walls, ceilings or other partitions, the openings remaining around the wiring system shall be made good to the appropriate standard of fire resistance.

This seal must be:

❏   Compatible with the wiring system.
❏   Permit thermal movement without reducing the quality of the seal.
❏   Easily removable for future extensions.
❏   Resist external influences to the same degree as the wiring system.

As far as possible, making-good should involve reconstruction using the original barrier material. This may be bricks and mortar or a concrete filling. Alternatively, purpose-made removable 'intumescent' barrier materials are available in the form of bags which may be packed into a hole.

Sealing the wiring system

Where a large trunking or conduit penetrates a wall or other structural fire barrier, the wiring system will need to be sealed internally to prevent the spread of fire and fumes.

Trunking manufacturers provide fire-block components which fit into the enclosure but permit the addition of further cables at a later date.

Smaller conduits and trunking need not be sealed provided that the wiring system is non-flame propagating. This relaxation applies to systems with a maximum internal cross sectional area of 710 mm^2 which approximates to a 32 mm diameter conduit.

Work in progress

527–03   It is important that sealing of fire barriers should be carried out as work progresses and before there is construction of false ceilings or decorative panelling. These may obscure sealing deficiencies. On a larger project the actual making good may be contracted to the builder but the electrical installer cannot escape the responsibility of certifying satisfactory completion.

In many instances electrical installation work is put in hand in an existing building. It is important not to put occupants at risk unnecessarily by leaving incomplete fire barriers. This would especially apply in hospitals, hotels and other residential buildings. The Regulations require that temporary sealing arrangements should be made.

Records

527–04 A record must be kept of work involving fire barriers for the information of the person carrying out the final verification. Entries in a working diary may be the best method of compliance. This system would also constitute valuable evidence at a later date should a dispute or disaster occur.

Hidden cables

There is an essential need to consider the physical protection of concealed sheathed cables which have no integral resistance to nail or screw penetration. It may be acceptable to assume that cables need no added protection if they are more than 50 mm beneath the wall, floor or ceiling surface. This is a matter of judgement by the installer.

Cables within the floor

522–06 Unprotected wiring must be run at a minimum depth of 50 mm beneath floorboards. It is usual to carry cables through holes drilled on the centre line of joists. Notching the top of joists is unacceptable both from electrical and structural viewpoints.

In some older properties it may be impossible to comply with the above restrictions. There may already be grooves made on previous occasions. In such circumstances cable may be taken near the surface provided that it has physical protection equivalent to heavy gauge steel conduit. Proprietary steel cover plates are available which give protection to electrical and plumbing services. Alternatively, purpose made plates may be used.

Cables above false ceilings

This topic is mentioned in other chapters.

❑ Cables may not be run less than 50 mm above the plasterboard beneath a battened concrete soffit.

❑ Either 50 mm battens are required or a cross-batten technique must be used. The latter gives clear routes for cables and is the preferable method of compliance.

Cables in walls

There is no practical alternative to the burying of cables within the plaster depth of walls in houses and commercial projects. It is usual to use metallic or plastic capping to retain cables in place and to avoid damage by a plasterer's float. This capping provides no physical protection against future penetration by a nail or masonry drill.

Unless enclosed in steel conduit or similar protection, cables running within 50 mm of the wall surface may only be routed within wiring zones.

- ❏ Cables may be installed within 150 mm of the top of the wall or a corner angle.
- ❏ Where a cable is connected to a point or accessory on the wall, cables may run horizontally or vertically to that position.
- ❏ Diagonal runs, no matter how short, are inadmissible.

Figure 14.1 illustrates the zones and shows how a supply to a cooker outlet should be routed.

Mechanically protected cables

Cables installed in steel conduit, or swa and MICS cables are considered to have adequate physical protection and the above restrictions do not apply.

Fire and smoke

523–01   It is probably true to say that no standard wiring system has advantages over any other in the prevention of a fire. A cable selected and installed in accordance with the Wiring Regulations is safe until engulfed in an external

App. 4   fire. The application of correct overcurrent principles should ensure disconnection of the circuit before a cable initiates a fire.

The essential fire resistant characteristics of a wiring system are primarily concerned with insulation, rather than conductor material. Two significant factors are:

- ❏ The contribution that combustible cable insulation and sheathing makes to fire, smoke and fumes, *and*,
- ❏ The continuity of supply given by the cable before disintegration.

The maximum operating temperature rating of a cable may be related to the above, but in most cases is of little consequence in an inferno.

PVC insulation

At normal working temperatures PVC has excellent electrical characteristics and resists chemical deterioration in most wet and dry conditions.

523–01   If an unplanned rise in temperature involves PVC overheating to the point of chemical decomposition, there are serious problems. Thick black clouds of

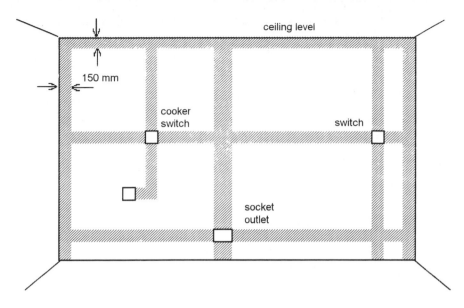

Figure 14.1 **Wiring zones in walls.**

choking acrid smoke arise which must be avoided at all costs. Apart from the suffocation consequences of burning and deprivation of oxygen, a major hazard to the victim is the emission of halogens of gaseous hydrochloric acid. At the very least this will cause a sore throat, and is quite likely to kill.

As a fire involving PVC develops, acidic fumes combine with water vapour and condense on surrounding surfaces. This will result in severe corrosion to both copper and steel. This will occur even with moderate overheating of cables, and the results are often seen after a fault on copper conductors and terminals. Any printed circuit boards in the vicinity of an overheated cable may suffer irreparable damage. For this reason halogen-free cables are often specified for mainframe computer suites. It must be remembered that, in the course of combustion, non-electrical PVC products will add to the fire, as will other flammable plastic materials, such as paint and furnishings, which are used in all buildings. The contribution from electrical products may be relatively small. In the case of plastic conduit and trunking, the qualities of unplasticized PVC introduce a favourable element of resistance to burning.

Polythene is no alternative

Polythene is more correctly known as *polyethylene*. This is not a suitable alternative or substitute for PVC. The material has excellent electrical characteristics but as an extruded *thermoplastic* material it has physical disadvantages.

❏ It is relatively soft, melts at low temperatures and has little 'scuff resistance'.
❏ Unless specially treated it will support combustion, and when burning produces flaming droplets.

Wherever thermoplastic polyethylene has been used for power wiring it has had to be protected by PVC or a similar sheathing.

Cross linked polyethylene

In recent years the cable industry has been able to extrude a *thermosetting* version of cross linked polyethylene which is known as XLPE and possesses useful characteristics:

523–01 ❏ XLPE will operate at higher temperatures than PVC.
❏ Higher operating temperatures mean less copper in cables where volt-drop is not a limiting factor.
❏ Flame retardant qualities are good.
❏ Virtually no hydrogen chloride is given off in a fire.

Steel wire armoured cables are now supplied as standard with XLPE insulation for sizes in excess of $16\ mm^2$. This has introduced an interesting variation in connection with the possible use of steel armouring as a circuit protective conductor. A thinner extruded XLPE insulation cover is applied to conductor cores than with a similarly rated PVC insulated cable. On most sizes this results in a smaller overall cable diameter and less steel wire armouring. The conclusion is that the armouring is inadequate for use as a cpc on many XLPE cables. The subject should be noted but is not relevant to this consideration of fire precautions.

XLPE compound is not as robust as PVC. For this reason the outer sheathing of standard steel wire armoured, XLPE insulated cable is usually PVC. A cable manufacturer should be consulted where a halogen-free sheathing is required.

Silicone rubber

Cables with this insulation are available in multicore form and most commonly used for fire alarm installations. Silicone rubber will disintegrate in a fire but the resultant ash is, to a certain extent, physically stable and remains as a good insulator.

Unfortunately silicone rubber does not have the same qualities for mechanical protection and therefore usually the outer sheathing is normally PVC. This limits the operating temperature to PVC levels and does not eliminate the emission of hydrogen chloride in a fire. This point should be noted in connection with computerized, addressable fire alarm controls which could suffer the consequences of acidic vapour corrosion.

To avoid this problem special low smoke, zero halogen (LS0H) sheathing is available. A single core, conduit version of silicone rubber cable has a braided glass oversheath to give the requisite mechanical protection. The absence of PVC reduces the problem of fumes.

Low smoke zero halogen (LS0H)

The days of simple single compound plastics are rapidly disappearing. Chemists are producing complex multi-polymers with engineered characteristics. LS0H comes in this classification.

There are various LS0H compounds which may differ to suit the intended application. In other words, LS0H is a description of a characteristic, not a material. The common factor with all LS0H cables is compliance with appropriate emission standards. Thus, one manufacturer can provide conduit cable with inner and outer composite layers giving specific insulation and fire resistance, and another manufacturer has managed to combine similar qualities in one layer.

The advantages of LS0H cable are with high operating temperatures and virtual elimination of smoke and halogen fumes.

Mineral insulated cables

Undoubtedly these cables are supreme for fire resistant qualities and will continue to function throughout the early stags of a conflagration. Eventually, of course, even copper melts.

523–01   The limiting factor for fire resistance will probably be at the cable termination.For special risk areas, high temperature terminations are advisable.

It is common practice to use PVC-sheathed MICS cable to give good protection against corrosion. This again introduces the problem of halogen release in a fire and advice should be sought from manufacturers regarding alternative sheathing.

Wiring systems and cable management

One of the principal reasons for selecting a particular wiring system will have been covered by the above explanation of fire resistant characteristics. Other reasons will be related to cost, appearance and functional performance. In larger building complexes the selection of wiring system will be designated as a cable management subject.

Emergency systems

In the following brief review of care with wiring systems, some of the cables mentioned will be used for fire alarm and emergency lighting circuits. Installers should be familiar with the appropriate British Standards before undertaking such work. A local fire or licensing authority may also have ideas upon cable selection, but the ultimate responsibility rests with the installation designer.

Care with wiring systems

In other chapters suggestions are given for the use of suitable cables on specific projects. Some explanation is given to the reasoning behind each recommendation. In practice, deciding factors will take into account safety and economy plus the external influences of environment, utilization and building construction.

A designer must make a final decision on the wiring system based upon the actual installation conditions. There is no perfect all purpose wiring system.

The following series of checklists cover some of the on-site practical measures that need to be considered with the installation of standard wiring systems. Both the installing electrician and commissioning inspector must take note of correct procedures and workmanship. In many cases, on smaller projects, total responsibility will rest with one person.

PVC insulated and sheathed cable

522–08 ❑ This cable is vulnerable to physical damage. Short lengths of protective PVC conduit or mini- trunking should be used if the cable is unavoidably routed at skirting level or similar positions.

526–03 ❑ The entire outer cable sheathing should be taken into fire resistant accessory or junction boxes.

543–03 ❑ A bare circuit protective conductor (cpc) must have the correct green/ yellow sleeving applied.

522–08 ❑ Cable runs must be correctly supported. A clip spacing of 300 mm is appropriate for most accessible locations. Particular care must be taken where cables are installed above metallic false ceiling grids.

PVC insulated conduit cables

521–07 ❑ Cables with insulation alone and no integral physical protection must always be enclosed in conduit or a continuous trunking system. Care must be taken at junctions and terminations to ensure that the protective enclosure is complete.

Steel conduit systems and trunking

543–03 ❏ Earth continuity along the conduit is the most important factor – even if an internal green-yellow cpc is used. Continuity through couplers and brass bushes must be permanent and reliable taking into account the proposed life of the installation. Completion documentation will virtually be certifying that continuity will still be adequate to handle fault currents in five or more years' time.

❏ Reliable continuity arrangements are essential with trunking or metallic skirting systems. The correct linkage should be used at joints and terminations.

❏ Protection against corrosion is essential and must take into account the
522–05 surrounding environment and conditions of use. Exposed threads or pipe-vice marks must be painted over to maintain the protective quality of the conduit. Cut ends of trunking may need protection.

❏ Conduits must be suitably supported. This especially applies to vertical drops to machinery in workshops. In these circumstances it may be preferable to convert to flexible cable for the ceiling to machine linkage.

Plastic conduit systems and trunking

The appearance of a plastic conduit or trunking system usually indicates whether sufficient care has been taken in the installation. There is no excuse for distorted or sagging conduits which result from inadequate fixings and lack of expansion facilities.

❏ A good general purpose rule is that conduit and trunking support is required at about 1 m intervals. Where ambient temperatures are high, or the installation is subject to solar radiation, spacing should be adjusted accordingly.

❏ Rigid PVC will expand and contract by about one millimetre per metre along the length of run in the normal temperature range that can be expected in the UK. To avoid problems, expansion couplers should be used at about 4 m intervals and saddles must allow for lateral movement.

❏ There are limitations in the weight that can be suspended from a PVC conduit box. Much depends upon the fixing of the box. Where a hot lighting fitting is connected directly to the conduit system, steel support clips should be used or a special heat resistant box.

Mineral insulated copper sheathed cables

❏ It is usual to use the copper sheathing as the cpc. This often causes problems regarding the adequacy of continuity at joints. Unless the installer is prepared to certify long term reliability at all joints, it is good practice to use pot seals with earthing tails.

❑ Voltage surges occur when inductive loads are switched. This sometimes causes a failure of the mineral insulation. It is advisable to fit surge diverters on cables supplying induction motors and fluorescent lighting installations.

Steel wire armoured cables

❑ Earthing continuity through the armouring gland at terminations must be effective, even if a separate cpc is used. A gland earthing ring should always be used with a cable linkage to the earthing terminal within a distribution board.

❑ The PVC outer cable sheathing is water resistant under normal conditions but may not be suitable for continuous immersion. The cable manufacturer should be consulted and a suitable sheathing specified.

Silicone insulated PVC sheathed cable

❑ Care must be taken when stripping insulation. Silicone rubber is much softer and has less mechanical strength than PVC. Protective ferrules should be applied when sheathing is removed.

Chapter 15

Completion

Procedures for the completion and commissioning of a project are as important as any of the construction activities. Electrical installation work is often not subject to inspection and testing by another person or third party. Self-certification is acceptable but does carry great responsibility.

With self-certification there is an absolute requirement for full inspection and test procedures. Everyone makes mistakes but when work is claimed to comply with safety regulations, the installer will be negligent if faults are not rectified.

This chapter explains the minimum completion procedures that are required by Health and Safety legislation, and the Wiring Regulations or BS 7671.

One would not expect to buy a car or computer which had not been tested. These products usually arrive with inspection certificates or labels which show that full procedures have been carried out. Every reputable wiring accessory used on an installation carries a mark which indicates certification to acceptable standards. In the case of mass-produced items these involve routine and random sampling.

A customer has the right to expect that the electrical installation is of the same standard as the quality components which are used. This product is only as good as the continuity of hidden connections, the use of correct cable sizes, and many of the other items that make up a complete installation.

Switching on to see that the light works is not a good enough test. No testing is acceptable until someone is prepared to put a signature to a document certifying safety. No occupier should be put at risk with an uncertified or untested installation. This point should be clearly explained to a customer who wants the system to operate before completion procedures have been carried out. It would not be expected with a car or aircraft. The most important aspect of an electrical installer's business should be tried and tested safety.

Labelling and documentation

The comparison with motoring continues. Where the purpose is not obvious, all the switches and controls on a car are marked to indicate their functions. A driver's manual is provided for the purchaser and a technical service manual is made available for future maintenance and repair.

Similar information and documentation is required for an electrical installation.

Specification and manual

IEE Guidance Notes suggest that each project should start with a specification. Ideas for specifications are given with each of the projects in this book.

At the completion of a job, information from the specification will form the basis of a user manual. In the case of a commercial project, Health and Safety legislation is concerned with the supply of suitable documentation. This composite manual should contain:

App. 6   ❐   Completion certificate as required by the Wiring Regulations,
  ❐   Schedule of circuits and test results,
  ❐   Advice regarding the use of equipment such as an mcb or rcd, and
  ❐   Manufacturers' information regarding equipment.

The complexity of the manual will be related to the size and complexity of the project. A simple house wiring contract will probably result in a manual containing the completion certificate plus some manufacturers' leaflets covering the protective devices at the consumer unit.

Regulations

The Wiring Regulations require that every installation shall be inspected and
711–01 tested both during erection and before being put into service. This covers the common situation where some parts of the installation become hidden from view as builders work progresses. It could be argued that inspection of inaccessible parts is more important than places where defects are visually obvious.

The person signing certification takes responsibility for all work, hidden and obvious. On larger projects a system of programmed inspections may be necessary. The subject is less complex where work is self-certified.

Completion certificate

The format for a certificate is given in the Wiring Regulations. Figs 15.1 and 15.2 illustrate the two parts of one version of the certificate produced for
App. 6 general use. Contractors approved by the following organizations will use certificates issued by these organizations (some specifiers insist on particular

COMPLETION AND INSPECTION CERTIFICATE

No. ....

DETAILS OF THE INSTALLATION:

Client: ....

Address: ....

....

....

EXTENT OF THE INSTALLATION:

....

....

....

See Notes inside pad cover for guidance
on compilation of this certificate

DESIGN

I/We being the person(s) responsible (as indicated by my/our signatures below) for the Design of the electrical installation, particulars of which are described overleaf, CERTIFY that the said work for which I/we have been responsible is to the best of my/our knowledge and belief in accordance with the Regulations for Electrical Installations published by the Institution of Electrical Engineers, 16th Edition, amended to (date) ....except for the departures, if any, stated in this Certificate.

The extent of liability of the signatory is limited to the work described above as the subject of this Certificate.

For the DESIGN of the installation:

Name (In Block Letters): ....

Position: ....

For and on behalf of ....

Address: ....

....

....

Signature: ....

Date: ....

CONSTRUCTION

I/We being the person(s) responsible (as indicated by my/our signatures below) for the Construction of the electrical installation, particulars of which are described overleaf, CERTIFY that the said work for which I/we have been responsible is to the best of my/our knowledge and belief in accordance with the Regulations for Electrical Installations published by the Institution of Electrical Engineers, 16th Edition, amended to (date) ....except for the departures, if any, stated in this Certificate.

The extent of the liability of the signatory is limited to the work described above as the subject of this Certificate.

For the CONSTRUCTION of the installation:

Name (In Block Letters): ....

Position: ....

For and on behalf of: ....

Address: ....

....

....

Signature: ....

Date: ....

INSPECTION AND TEST

I/We being the person(s) responsible (as indicated by my/our signatures below) for the Inspection and Test of the electrical installation, particulars of which are described overleaf, CERTIFY that the said work for which I/we have been responsible is to the best of my/our knowledge and belief in accordance with the Regulations for Electrical Installations published by the Institution of Electrical Engineers, 16th Edition, amended on (date) ....except for departures, if any, stated in this Certificate.

The extent of liability of the signatory is limited to the work described above as the subject of this Certificate.

For the INSPECTION AND TEST of the installation:

Name (In Block Letters): ....

Position: ....

For and on behalf of: ....

Address: ....

....

I RECOMMEND that this installation be further inspected and tested after an interval of not more than ...... years months

Signature: ....

Date: ....

Acknowledgement: This Certificate is based on the model
contained in the 16th Edition IEE Wiring Regulations.

Published by UNEEDA PUBLICATIONS 1991

Figure 15.1 Completion certificate requires three signatures. (Reproduced by kind permission of UNEEDA Publications.)

PARTICULARS OF THE INSTALLATION

(Delete or complete items as appropriate)

Type of Installation: New/alteration/addition/to existing installation

UNEEDA

Type of Earthing (312-03):   TN-C   TN-S   TN-C-S   TT   IT
(Tick box)   ☐   ☐   ☐   ☐   ☐

Earth Electrode: Resistance .... ohms   Instrument make and number ....

For TT or IT Systems ⎡   Method of Measurement ....
                    ⎣   Type (542-02-01) and Location ....

Characteristics of the supply at the origin of the installation (313-01):

Nominal voltage .... volts   Number of phases ....

Frequency .... Hz

	Ascertained (Tick box)	Determined (Tick box)	Measured (State instrument make and number)
Prospective short-circuit current (I_p) .... KA			
Earth fault loop impedance (Z_E) .... ohms			

Maximum demand .... A per phase

Overcurrent protective device: Type BS ....   Rating.........A

Main switch or circuit breaker (460-01-02): Type BS ....   Rating.........A   No of poles ....

(if an r.c.d., rated residual operating current I_{Δ_n} .... mA.)

Method of protection against indirect contact:

1.   Earthed equipotential bonding and automatic disconnection of supply ☐   (Tick box)

2.   Other ☐   (Describe) ....

Main equipotential bonding conductors (413-02-01/02, 547-02-01): Size .... mm^2

Details of departures (if any) from the Wiring Regulations (120-04, 120-05):

Comments on existing installation, where applicable (743-01-01):

See accompanying Test Schedule(s) No(s) ....for test results

See accompanying Completion Schedule(s) No(s) ....
for details of installed equipment and results of visual inspection.

Note: This Completion Certificate does not cover portable appliances or apparatus connected to socket outlets.
UNEEDA Publications Appliance Test Record may be used for this purpose.

Acknowledgement: This Certificate is based on the model contained in the 16th Edition IEE Wiring Regulations.

Published by UNEEDA PUBLICATIONS 1991

Figure 15.2   **Installation particulars shown on completion certificate. (Reproduced by kind permission of UNEEDA Publications.)**

forms; the form may be supplemented with inspection reports for special areas): ECA (Electrical Contractors' Association); ECAS (ECA of Scotland); and NICEIC (National Inspection Council for Electrical Installation Contractors).

Signatories

Three signatures are required on the completion certificate:

Designer

This person may be a consultant, a local authority engineer or the installer. The designer will have made electrotechnical judgements and perhaps calculations. Generally architects and builders do not design electrical installations. They may give instructions upon the type and location of equipment and cable runs, but this is building design, not electrical design as required by the certificate.

An installer who uses rule-of-thumb methods or adopts published designs, or copies other designers' work must take responsibility for the project design. This implies that the installer must have the competence to design the work although an acceptable short-cut has been taken.

In some cases a consultant may pay an installer to verify the design and take responsibility for the certification.

Installer

The designer cannot be held responsible for the way in which an installer follows the specification. The installer therefore certifies that there is compliance with the Wiring Regulations as far as on-site installation work is concerned. This will include the selection of appropriate materials to match the specification or to comply with Regulations. For example, wherever appropriate, all materials should carry a British Standard mark or other acceptable product certification.

This subject is summed up in the fundamental requirement in the Regulations stating that good workmanship and proper materials shall be used.

It is in this part of the project that inspection and testing may be necessary as the work progresses. It may be that the installation is intended to be progressively energized and put into service on a phased basis. In this instance interim certification may be necessary to confirm that the installation is safe and fit for use. The Regulations offer no relaxation in safety standards for temporary work. It may even be considered that a higher standard of safety is necessary for incomplete installations used by people who will be unfamiliar with the system.

Inspection and test

711–01 This may be carried out by a specialized operator, or the installer, or perhaps the design consultant. The inspector must be in possession of design data and the specification. Testing is carried out to ensure that the completed project complies with the Wiring Regulations. Some aspects of the inspection may have been carried out during the progress of the work. This particularly applies to making good holes made in fire barrier walls for cable routes.

Full testing must be carried out before the system or part of the system is energized and handed over to the client. Future users should not be put at risk by incomplete work.

Extensions and alterations

All work is subject to the full test requirements. Small jobs and alterations must be certified in a similar manner although in some cases a short form certificate may be acceptable to the client.

Limits of responsibility

A signatory on a completion certificate can only be responsible for work for which there is a contractual arrangement. If, for example, the client only requires a house extension to be wired, the installer cannot be held responsible for the existing installation in the older part of the building unless this is part of the agreement. The situation is illustrated clearly if one considers the consequences of installing an additional socket outlet in a large multi-storey office complex. There is no way that the installer is going to certify the quality of the electrical system in the rest of the building.

The installer carries the obligation to ensure that the new work complies with current regulations and that down-stream equipment supplying the addition or extension is adequate for the additional load.

There is a further requirement which may have significant consequences. That is to ensure that the earthing arrangements for the extension are satisfactory. The problem is that each socket outlet and service connection relies upon main earthing and bonding for safety. It is suggested that this matter is investigated before the contract is accepted.

In the case of small domestic additions or alterations, for example the installation of a shower, the importance of correct bonding cannot be over-stressed. If there is any doubt about the old installation, the best solution is to install a new consumer unit for the shower and to establish this as a separate installation with the relevant earthing and bonding.

Regardless of the above, an installer, as an electrical expert, has a moral and statutory duty to bring to the attention of the occupier any obvious dangers in the old installation. In extreme cases this might best be covered by a written recommendation stating that the installation should have a full inspection either straightaway or within a stated period of time.

Deviations and departures

There may be circumstances in which some aspects of the installation do not strictly accord with the Wiring Regulations. The fact that there is a space on the form to record such departures implies that some of these may be acceptable.

New ideas and methods

120–04 In order to accommodate new ideas and methods, departures from the Regulations are acceptable provided they do not result in lower safety standards. This does not encompass situations where a client insists upon features that are in conflict with the Regulations. An example may be a request for a bathroom socket outlet for a hair drier. In these circumstances the installer may not deviate from the Wiring Regulations and must refuse to carry out the work. In the case of the hair drier it would be diplomatic to suggest alternative ideas for a fixed appliance.

Customer responsibility

There may be other, more reasonable requests by the client which may be accommodated with safety. For example, a householder may refuse to have any rcd protection on socket outlets. An assessment of the situation might indicate that, in the absence of any suitable appliances, no sockets may reasonably be expected to supply portable equipment for use out of doors. A decision must be reached as to whether this constitutes a departure from the Wiring Regulations which are intended to cover all possible future use of the installation and not just today's conditions with the present occupier. If after full discussions the client still refuses the rcd, it is suggested that this is shown as a departure on the completion certificate, and the client be asked to initial the note.

Certainly an installer cannot plan for such departures from the Regulations as a way of quoting for a cheaper job.

Particulars of the installation

One whole page of the completion certificate is intended to carry a summarized technical report on the installation (Fig. 15.2). Most questions are easily answered and in some cases may be deleted where they are not relevant to the particular premises.

Some topics that raise queries are:

❏ *Prospective fault current or prospective short-circuit current*

313–01   This is the maximum instantaneous current that will flow under short-circuit conditions, either phase to neutral or phase to earth. Instrumentation is available to measure this condition. On a single-phase 240 V pme system where neutral and earth are interconnected at the electricity company's terminals:

$$\text{pfc} = \frac{240}{Z_e \times 1000} \text{ kA}$$

where Z_e is the earth fault loop impedance in ohms at the consumer unit.

❏ *Overcurrent protective device*

This is the electricity company's fuse. On a pme domestic installation it will probably be BS 88 or BS 1361: 100 A rating.

❏ *Main switch*

Details are shown on the front of the consumer unit or main switch.

❏ *Rated rcd current*

Where applicable, this will be 30 mA or 100 mA, etc.

❏ *Method of protection*

413–02   On the public supply this will be EEBADS. Alternatives would only apply to special unearthed systems, etc.

Installation and test schedules

A completed installation schedule should accompany the completion certificate with results of tests. There is no laid down format and all items may be combined on one sheet. Figure 15.3 illustrates two sheets covering separately the schedule of equipment and test results. These relate to circuitry at one distribution board. For large schemes, multiple copies will be required or purpose-made documentation produced.

Inspection procedures

Sec. 712 The Wiring Regulations detail at length the inspection checks that must be carried out before testing. In the IEE Guidance Notes the list is expanded to some 200 items that, in many cases, state the obvious. The idea is that *every* aspect of an installation should be viewed with a critical eye to ensure that both the detail and spirit of the Regulations have been met.

Figure 15.3 It is important a full inspection check list be produced. (Reproduced by kind permission of UNEEDA Publications.)

Testing

Sec. 713 The Wiring Regulations state test requirements, but not test procedures. Suggestions for suitable test methods are given in an IEE Guidance Note but alternative methods may be used provided that the results are equally effective.

Testing ideas shown here summarize the very minimum routine that may be applied to a simple installation. For detailed information refer to the IEE Guidance Note.

It is important that the tests are carried out in the correct sequence, using reputable and reliable equipment. Instrument manufacturers' instructions and full safety procedures should be followed.

❑   Where testing requires the exposure of live parts, the inspector should always be accompanied.

❑   Low cost, uncertified multimeters are not suitable for any of the tests required by the Wiring Regulations.

❑   The use of home-made test lamps or test leads would probably be a breach of Electricity at Work Regulations.

❑   Where tests are made on the supply, fused probes must be used.

❑   Instruments must be regularly calibrated. Where re-calibration indicates that faulty readings have previously been given, all earlier tests are suspect.

It is a good idea to keep some local test reference arrangements that can be applied weekly. A 'standard' coil of cable can be used to check the accuracy of a continuity scale and a particular convenient socket outlet used to confirm the integrity of the earth loop impedance tester.

Results of testing should always be recorded, otherwise when disaster strikes there is no proof that testing was ever carried out. Remember that all documentation may be required by a court of law.

Continuity testing

These tests are all carried out before the supply is connected. If at any time there is found to be a test failure, all previous tests involving the particular conductors must be repeated.

Polarity

The opportunity should be taken to check polarity when conducting the following continuity tests. The polarity requirement may be ticked off on the testing schedule after the relevant check.

Continuity of protective conductors

Figure 15.4 Illustrates a suitable method. Each cpc is connected in series with the corresponding phase conductor. The result gives the resistance of the fault path from an outlet or appliance to the distribution board. Readings must be supplemented by visual checks on terminations. Ideally this test can be carried out before final connection to the consumer unit.

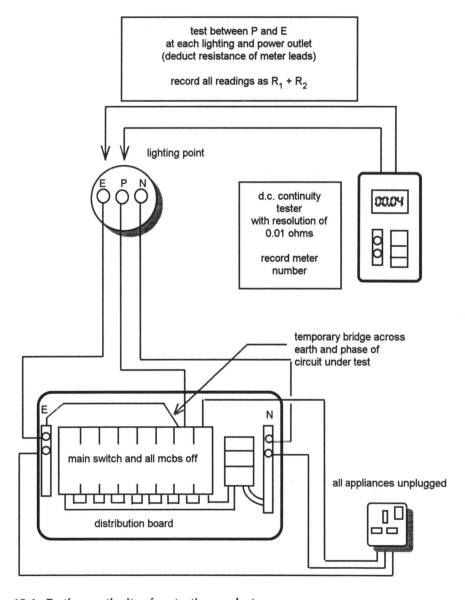

test between P and E
at each lighting and power outlet
(deduct resistance of meter leads)

record all readings as $R_1 + R_2$

lighting point

E P N

d.c. continuity
tester
with resolution of
0.01 ohms

record meter
number

temporary bridge across
earth and phase of
circuit under test

E

N

main switch and all mcbs off

all appliances unplugged

distribution board

Figure 15.4   **Testing continuity of protective conductors.**

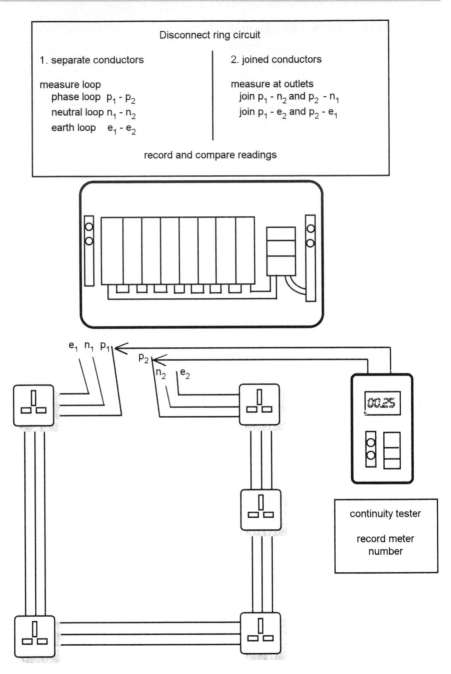

Figure 15.5 **Disconnect and separate ring circuit conductors to test for continuity.**

The test does not necessarily show the quality and long-term reliability of a joint, especially where corrosion may occur. Where a circuit relies upon continuity through steel conduit or the connections of conduit-type fittings to steel switchgear, testing with an earth loop impedance instrument is more effective.

Continuity of ring circuit conductors

713–03 Figure 15.5 shows a suitable arrangement. It may be convenient to carry out this test at the same time as each ring circuit is connected into the distribution board. This will ensure that there is no subsequent cross-connection of circuits.

Two sets of readings are taken.

Test no.1

The first set of readings will give a comparative resistance of cable loops. If one loop shows a noticeable difference to the other two loops, a check should be made of joints around the circuit. To make this test effective it will be necessary to compensate readings for differences in conductor sizes, e.g. a 1.5 mm^2 circuit protective conductor will have 0.6 times the resistance of a 2.5 mm^2 phase or neutral core.

Test no.2

This part of the test involves joining two loops in series. Readings are then taken at socket outlets and all should give identical results.

If all connections are correct with good joints, approximate comparative readings, as shown in Table 15.1, should appear.

Table 15.1 **Comparative earth continuity readings.**		
Test no. 1		*Test no. 2*
Ohms		
$p_1 \rightarrow p_2$	$=$	2 $(p \rightarrow n)$ at socket outlets
$(p_1 \rightarrow p_2) + (e_1 \rightarrow e_2)$	$=$	4 $(p \rightarrow e)$ at socket outlets

Insulation resistance

713–04 This test is carried out before the supply is connected.

All lamps should be removed and control equipment disconnected. This is particularly important if delicate transistorized devices are involved. Tests as shown in Fig. 15.6 should be applied, using a 500 V insulation tester. The reading across the whole system must be greater than 0.5 megohms but it is more usual to achieve full scale readings on each circuit. Inconsistent readings

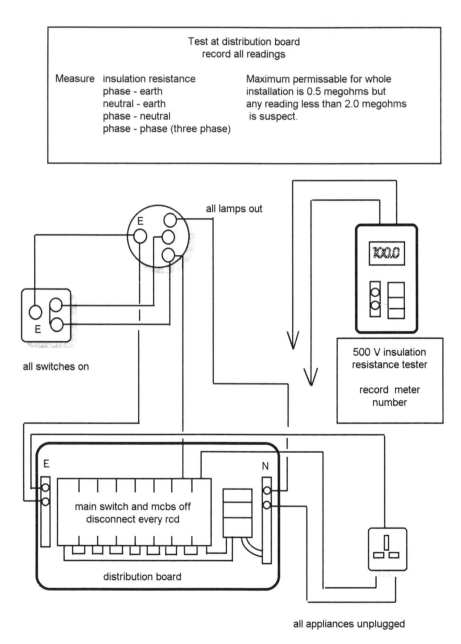

Figure 15.6   **Disconnect electronic circuitry before testing insulation resistance.**

or results midway across the meter scale usually indicate localized dampness or cable damage, for example by a nail. The cause of low readings should always be investigated.

Earth loop impedance

These tests are conducted after the system has been energized. They are possibly the most significant for future safety. If carefully interpreted, results will show:

713–10

- ◻ The adequacy of earthing,
- ◻ Poor connections,
- ◻ Cross polarity at the supply or anywhere in the installation,
- ◻ Continuity of ring circuit conductors,
- ◻ Correct positioning of switches,
- ◻ Cross connection of ring circuits.

The earth loop impedance test instrument injects a significant voltage on to the system. Care must be taken to ensure the safety of users of the installation or persons in contact with earthy metalwork whilst tests are in progress.

Any circuit found to be faulty should be immediately de-energized. After rectification, a repeat of all testing is necessary on the particular circuit and any related circuits.

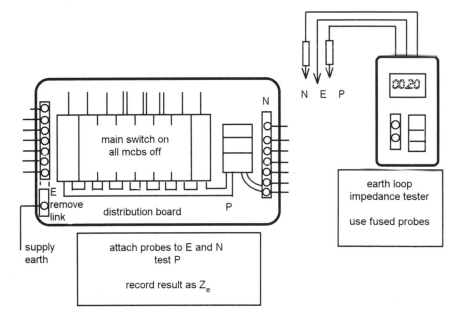

Figure 15.7 **Testing the external earth loop impedance.**

Supply impedance Z_e

The main earth will be taken from the supply company's earthing terminal or an earth electrode.

Disconnect this main earth from all bonding and circuit protective conductors for the test. Note that this leaves the installation unearthed, therefore all circuit-breakers should be switched off. Figure 15.7 shows the test method.

Table 15.2 indicates maximum readings for TN-S and TN-C-S installations. TT earth electrode resistances are covered in Chapter 10.

Earth loop impedance at outlets Z_s

Tests should be taken at every lighting position, socket outlet and equipment connection. Figure 15.8 shows the method of testing. The maximum reading on each circuit should be recorded. These are based on the maximum permissible Z_s impedances under fault conditions and multiplied by 0.7 to take into account the fact that tests are made cold.

Note that erratic results indicate poor connections. A steadily increasing reading around a ring circuit usually indicates an open circuit condition.

Table 15.2 **Maximum acceptable earth loop impedances adjusted for fault conditions.**

	Max. measured earth loop impedance (ohms)							
TN-S at main earthing terminal: 0.8 TN-C-S at pme terminals: 0.35								
mcb	*Current rating* (A)							
	5	6	10	15	16	20	30	32
Type 1 Z_s max. ohms	8.4		4.2	2.8		2.1	1.4	
Type 2 Z_s max. ohms	4.8		2.4	1.6		1.2	0.8	
Type 3 Z_s max. ohms	3.4		1.7	1.1		0.8	0.6	
Type B Z_s max. ohms		5.6	3.4		2.1	1.7		1.0
Type C Z_s max. ohms		2.8	1.7		1.0	0.8		0.5

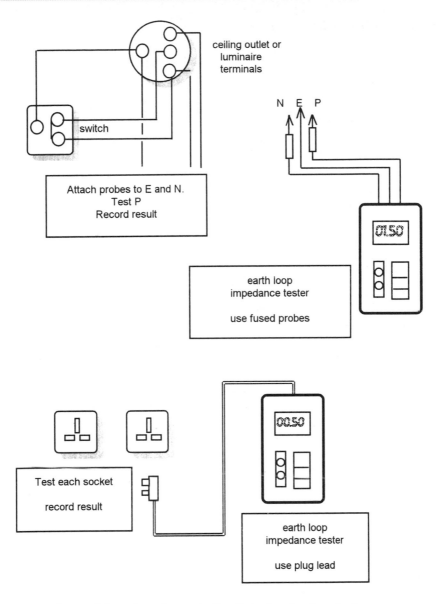

Figure 15.8 **Testing earth loop impedance at all outlets also checks polarity.**

Operation of residual current devices

Where an rcd protects socket outlets, a simple check is with the tester plugged into a socket. All other appliances must be unplugged to avoid the effect of equipment leakages.

If an rcd protects a whole installation or a circuit without sockets, the test should be carried out with clip-on probes at the rcd terminals.

Test procedures are as shown in Table 15.3. After carrying out these tests the operation of the test button on the device should be checked. This button

Table 15.3 Test procedures for standard residual current devices.			
Test	Tester setting	Pass result	Conclusions
'Instantaneous' residual current device			
1	50% of rating	rcd should not trip	There is no excess standing leakage on the circuit and rcd is not over sensitive
2	100% of rating	rcd should trip in less than 200 ms	Correct fault operation
3	150 mA	30 mA rcd should trip in less than 40 ms	Correct shock protection
Time delay residual current device			
1	50 % of rating	rcd should not trip	There is no excess standing leakage on the circuit and rcd is not over sensitive
2	100% of rating	rcd should trip in *not less* than 50% of rated time delay + 200 ms and *not more* than 100% of rated time delay + 200 ms	Correct fault operation and adequate time delay for discrimination

proves the internal operation of the device and not the external circuit. This test alone cannot be accepted to verify correct operation of the rcd to clear a fault.

Chapter 16

Advanced Ideas

This book is not intended to be a training manual. It gives ideas and explanations for simple installations with different features. The emphasis is on safety regulations and project management rather than technology. In practice, no project will match any of the schemes illustrated but by taking note of special requirements, the designs may be adapted.

Electricity at Work Regulations

These Regulations were published in 1989. Every installation contractor and designer should possess a copy of the H & SE Memorandum of Guidance. This gives a definitive interpretation of an important statutory safety document.

The EAW Regulations are concerned with electrical systems, equipment and the use of electricity in the workplace. All situations are covered, including private and public establishments. Probably domestic housing is the only area excluded for the fixed installation but the Regulations would apply whilst work is in progress. Housing is of course covered by the IEE Wiring Regulations and contractually every installer should comply with this standard for all work.

Competence

An important matter for the installation industry is the requirement to employ competent persons. Any person engaged on electrical work must possess such knowledge or experience, or is under such degree of supervision as may be appropriate having regard to the nature of the work. Electrical installers must appreciate their personal limits of competence. They must not undertake projects without having a full understanding of the potential hazards.

Small simple installations involve basic design methods with minimal technology. Rule-of-thumb estimates for cable sizing and earthing practice will be adequate. Once a scheme moves into the realms of heavy current and large protective devices, there is a need for greater caution. Tremendous energy resources are being handled by electrical equipment. Incorrect assumptions on overcurrent protection may have disastrous results. A short circuit may cause extensive overheating and fire. There is also the prospect of flashovers and electromagnetic stress between conductors which produce explosive forces. The consequences of an electrical accident are often horrific.

An electrical installation in a large building complex is not just a multiple of several small jobs. This chapter gives guidance upon electrical design factors that need to be taken into account. Readers should know their own limitations and seek training or experience where necessary.

The following information is intended to be helpful to designers and installers who wish to extend their activities in bigger projects and more complex techniques. Cautionary advice is given upon venturing into unfamiliar commercial situations.

Live working

H & SE Regulations indicate that live working must always be avoided unless it is unreasonable in all the circumstances for the system to be dead, and suitable precautions have been taken to prevent injury.

Inconvenience to the occupier or expediency by the installer are not reasonable grounds. Live working should only be undertaken after consultation with the client and the issue of a Permit to Work certificate. It is essential that a person working on live equipment should be accompanied by another person who understands the work in hand, the risks and emergency procedures. All electrical operatives should have a knowledge of first aid resuscitation.

Earthing

There have been many technical papers and academic conferences on the study of earthing technology. This is not a simple subject. There is a continuous debate, for example, upon one topic alone: the efficacy of protective multiple earthing. This book can only advise upon primary safety limits.

The earth is not a great sponge that soaks up unwanted leakage current. It is a general-purpose conductor with variable characteristics which everyone connects up to either deliberately or by accident. By its inconsistent nature, the earth will transmit current indiscriminately in various directions. The aim of an installation designer is to provide safe and controlled access to this common earthing system.

Chapter 7 explained the principles of earth resistance and voltage gradients. Equipotential bonding will handle variations in voltage on a small site or between houses.

Prospective fault current

Circuit protection for lighting on a 5 A fuse or circuit breaker will operate with an earth loop impedance of about 10 ohms. A lower resistance is necessary for a 30 A mcb. The fault path impedance must be less than 1 ohm.

This is the minimum circuit and earth resistance that will pass sufficient current to operate an mcb quickly enough to disconnect before dangerous energy is released.

With a 1000 A fuse, earth continuity must be in the region of a few hundredths of an ohm, depending upon the characteristics of the device and nature of the fault path. Even this low resistance will be inadequate with the prospect of more than 20 kA fault current on the supply company's low-voltage network. This is why the Regulations specify large cables for bonding on large installations.

Earthing calculations and measurements operate on a different order of magnitude for larger projects. Frequent calibration of instruments is important. Earthing and bonding conductors must be carefully routed with positive connections. Earth electrodes should only be installed by specialists. The local electricity company is usually the best source of advice.

A main earthing or bonding connection must never be casually disconnected, even on a small installation with the supply switched off. The green/yellow cable may be carrying network current.

Similarly, earth and neutral should never be interconnected within an installation. The supply company's pme connection is not a random operation. It is a calculated procedure.

Study starts with the Wiring Regulations and the adiabatic equation for sizing protective conductors.

Overcurrent protection

Many of the principles for protection against fire and shock hazards rely upon the installation of the correct overcurrent protective devices. Chapter 4 is an introduction to the subject.

This book deals exclusively with miniature circuit-breakers which are comfortably backed up with the supply company's main fuse. The selection of overcurrent devices should never be lightly undertaken. Consideration should be given to the reasons for a fuse blowing or mcb tripping out. Sometimes it is found that a circuit is not operating correctly with no obvious faults. It is probably the arithmetic that has gone wrong.

A fuse or mcb should be selected to protect the down-stream cable and should not be uprated without good reason. A change in fusing may permit overloading and deterioration of the cable which, at the very least, could be expensive.

Heavy current devices

When a designer moves into the realms of large system protection, rule-of-thumb judgements are not applicable.Once again there is a move into higher technology and calculations are essential.

Some items of equipment used on smaller installations have characteristics that require careful selection of overcurrent protection devices. Inductive loads, such as discharge lighting and larger boiler pump motors, do not follow elementary ohm's law calculations. The problem is magnified in large buildings with air conditioning plant and complex computerized machinery.

Problems are now coming to light with balanced three-phase loads on cables in high-tech offices. The supply 50 Hz waveform is being so polluted by assymmetrical high frequency signals that a neutral conductor sometimes carries more current than the associated phase wires. The days of providing a half-size neutral have long disappeared and the opposite may soon apply.

An installer should take care if a system contains unusual equipment or an installation has to be planned for speculative building operation.

Discrimination

Lack of discrimination for overcurrent protective devices may cause nuisance problems. The occupier of a large building will not appreciate a whole section of the works shutting down to clear a trivial local fault. Everyone can quote cases of inconvenience caused when sub-circuit fuses are by-passed. Careful planning in the selection of overcurrent devices can help to eliminate this condition, especially in places where continuity of supply is important.

Advisory services

The best source of guidance for overcurrent protection will be obtained from circuit-breaker manufacturers. Always read the technical part of the catalogue and look out for suppliers' training courses. Cable manufacturers will advise on I^2t characteristics of conductors. These services are all superior to unqualified decisions made after a chat with colleagues at the wholesaler's counter.

Isolation and switching

Chapter 11 indicated that decisions upon switching for safety are not always clear cut. Some devices need to be operated on load and this may cause problems with highly inductive equipment. The selection of a disconnection device can only be made in conjunction with a knowledge of the load in question. Once again manufacturers of equipment will give good advice.

Cable management

This is yet another developing topic with increasing complexity of services. A forward-looking installer will need to become familiar with wiring systems for power and communication cables. The subject of electromagnetic compatibility is important and may be offset by the general use of fibre optical systems.

Energy management is a term now used to replace heating and ventilation. Interactive controls establish a programmed environment. The wider concept of whole building management is being introduced not just for commercial buildings, but also in the domestic situation. It has already been suggested that new housing should be wired with circuitry that can handle future data transmission signals.

Standards and Regulations

As with those having expertise in any field, electrical installers are expected to answer customers' questions upon all aspects of electrotechnology. It is essential to keep up to date with the Wiring Regulations and statutory Health and Safety documentation. British Standards are written for the benefit of product users and are continually updated and harmonized. Codes of practice are readable documents which offer considerable help when planning installations in special locations.

The trade press usually gives news of amendments to publications but it is virtually impossible to maintain a personal library of all current standards and codes of practice. With the prospect of continual changes, it is not always wise for the smaller contractor to purchase documentation. Possession of an out of date standard may be worse than having no standard at all. Major county libraries now carry the whole range of current UK and international standards on compact discs. Access is easy and time spent in research will be invaluable.

Index